LIFE IN SPACE

Life in Space

ASTROBIOLOGY FOR EVERYONE

Lucas John Mix

HARVARD UNIVERSITY PRESS

Cambridge, Massachusetts

London, England

2009

Copyright © 2009 by the President and Fellows of Harvard College
All rights reserved
Printed in the United States of America

Library of Congress Cataloging-in-Publication Data
Mix, Lucas John.
Life in space : astrobiology for everyone / Lucas John Mix.
p.   cm.
Includes bibliographical references and index.
ISBN 978–0–674–03321–4 (alk. paper)
1. Exobiology.   I. Title.
QH326.M59   2009
576.8′39 — dc22   2008041260

To Gerald Soffen,

for putting my head in the clouds,

and to David Haig,

for keeping my feet on the ground as often as possible

Contents

LIFE IN SPACE

1

Caught Up in Life

Earth is a proper noun, a definite and individual place. Living close to the planet, we view it as something more generic: the ground beneath us, the soil, the background. Science has begun to paint a picture, however, of a more particular place. We inhabit one planet among many, and yet a tremendously interesting planet. Life happens here, and we as living things participate in it. Earth is a living planet.

This book deals with the growing field of astrobiology, the study of life in space. As we begin to explore the Solar System and the wider universe, we must ask ourselves what makes Earth unique. Is Earth unique? We want to know about our place in the cosmos and how we fit in. We want to know how we came to be. We want to know if we are alone. But our search for understanding only leads to more questions—ones that turn out to be very complex indeed. What is life? How do we recognize it? How does it work?

Various fields of science have started to give us important insights into these questions. Biology tells us things about life and how living things relate to one another. Astronomy and geology help us understand how stars and planets form. Chemistry teaches us about the fundamental stuff from which life has been made and how it cycles through stars, planets, and living beings—even us. Near the end of the twentieth century, many scientists came to the conclusion that all of these fields needed to communicate before we could come to a deeper knowledge of life. Astrobiology attempts to do just this. It incorporates insights from a wide variety of scientific disciplines to create a coherent picture of life. If life is something that happens to planets, if it is a phenomenon not just of our place but of the universe, then only a conversation among many disciplines will produce a coherent picture of it.

Even with scientific collaboration, we face a formidable barrier: We have only one example of life. It may not seem so at first, but if you think about it, life on Earth appears to be integrated. We can't break off bits and study them in isolation. Before long, they cease to be life. I am not saying life on Earth cannot be reduced to smaller components, only that we have not really managed to do so yet. Ecology displays the interconnectedness of living things. Evolution suggests that all organisms on Earth are related. In some sense, all life on Earth can be thought of as a single, holistic living process.

Much of this book will deal with the surprising number of features shared by all known living things. All rely on the same physical laws and entropy. All are built out of carbon molecules interacting in solutions of water. All manipulate electrons to store and use energy in a surprisingly small number of ways. All life on Earth uses the same set of chemical building blocks to get things done. Analysis of these building blocks gives us good reason to believe that these similarities occur because all life shares a common ancestor. At some point in Earth's past, some thing emerged from the chemical soup and started having children. As far as we know (and there is so much left to learn), all organisms now living descended from that one little . . . something.

If we imagined ourselves to be silicon-based tourists from Titan, we might be horribly disappointed to discover that life on Earth was all pretty much the same. No creativity at all; no real variety. As a biologist I was equally disappointed. How can you study life if you only have one sample? If you want to know about anything in a general sort of way, you need to work with multiple examples. The more I learned about biology—the study of life—the more I realized that it was really the study of one sort of life—life on Earth. And that summarizes how I became involved in astrobiology. I wanted to know about more than our little piece of life. I wanted to know about life as a grand, universal, wonderful phenomenon in the universe. I wanted the big picture.

So is astrobiology the "big picture"—life writ large? Yes and no. It certainly deals with life writ large, but it does so in a very specific way. I promise; I'm getting there as fast as I can. We were talking about life on Earth.

Astrobiologists use the word "terrean" to describe things about our planet in particular. The word "terrestrial" was already taken by the planetary scientists and the botanists.[1] If you are a fan of science fiction, you may be more familiar with the word "terran." It means the same thing, and many astrobiologists have started using this spelling instead.

I am not going to knock life on Earth. It is pretty spectacular and I have devoted many years to the study of it. I can also say that it has been particularly good to me. After all, if not for life on Earth, I wouldn't be here. I wouldn't have anything to eat, and, being quite hungry, I wouldn't be inclined to write books about the nature of life. And that, in a nutshell, is our second barrier to understanding. We are inextricably caught up in life on Earth. It can be tremendously difficult to be unbiased.

One of the most amusing and profound consequences of this second barrier arises when we ask about the probability of life existing in the universe. I like to say that we have only visited one planet and it has life on it, so, as far as we know, every planet bears life. Statisticians call it a "sampling problem" when the available examples are biased with regard to whatever you want to study. We have a sampling problem. The most available planet has abundant life—us. What does this say about our ability to judge the conditions of our existence? What about the likelihood of our appearing at all? Elements of this second barrier have been called the anthropic principle and the Goldilocks principle, both of which will be covered in Chapter 5.

What Astrobiology Is

Astrobiology is the scientific study of life in space. It happens when you put together what astronomy, physics, planetary science, geology, chemistry, biology, and a host of other disciplines have to say about life and try to make a single narrative. Life happened to the universe in the context of Earth. It affects, or perhaps infects, whole planets and cannot be understood at a smaller scale—at least not yet.

For me, the most interesting parts of astrobiology address the dividing lines between disciplines. Where does chemistry end and biology begin? The two fields flow into each other, of course, but traditionally chemists stop at some point. They hand things over to biochemists, who in turn only do so much before handing things over to physiologists, organismic biologists, ecologists, and so on. Astrobiology attempts to bridge the gaps between different fields, to pry into precisely where those areas of study fit together and overlap. As astrobiology grows, a new vocabulary will be built and a new way of looking at life will emerge. It will not be the life of individual organisms or even of ecosystems, but the life of planets and, hopefully, life in the universe. This book introduces the foundations upon which such an understanding may grow. The edifice is not yet in place, but we are beginning to see glimpses of the whole.

What Astrobiology Is Not

Having started with this understanding of astrobiology, I should speak about a couple of things that are not included. Astrobiology is not the study of alien life. To my knowledge, no scientist on Earth currently studies alien life, or xenobiology. Many scientists research the extent of life and how it might be possible elsewhere, but our data come from terrean life and exploration. Another word you may hear from time to time is exobiology, the study of life outside—beyond Earth. Sometimes this means xenobiology and sometimes it means the study of how space environments affect terran organisms. The latter has produced some fascinating data in the last few decades as we send humans and other organisms into orbit.

I would also say that astrobiology is not the study of the origin of life. That would presume once again that we have access to more data than we actually do. Astrobiologists often do look at existing life and extrapolate backward. Current theories suggest that long ago there was a common ancestor to all extant life on Earth. This organism can be inferred by extension from Darwin's realization—supported by mountains of subsequent observations—that two modern, coexisting species were once only one. Skipping over the subtle complexities of phylogenetics, if we start collapsing species into each other, we can see that, perhaps, a single species predated them all. Getting from one to many, however, is vastly easier than getting from zero to one. (If you're interested in this idea, it's a favorite philosophical playground of mathematicians.)

Biologists have yet to find a way to get life from nonlife. A very well-known scientific theory, "spontaneous generation," floated around in the seventeenth, eighteenth, and nineteenth centuries. It was commonly thought that maggots and flies arose spontaneously from decaying plants and meat. But in 1668 Francesco Reddi noted that eggs were present in these rotting materials. Over the next two centuries, experiments showed that insects laid the eggs, which later hatched to form new insects. After Louis Pasteur's experiments around 1862, biologists formed a general consensus that spontaneous generation does not occur. When no insects were allowed near, no new insects grew. Spontaneous generation was considered disproved. Careful experiments show that when life is excluded from an environment, new life does not emerge. Life does not arise from nonlife.

Astrobiologists would like to have a better understanding of how life came into being on Earth. Several labs exist that pry into the fundamental structures of organisms. They make theories about the biochemistry required for life and how a purely chemical world may have changed into a biological world. They have uncovered valuable insights into how life operates and what may be essential. Still, we cannot reproduce or confidently model the transition from nonlife to life. We simply do not know how to reconstruct this single, monumental event.

Some astrobiologists make much stronger claims about the origin of life. Although unable to reconstruct the origin of life, scientists are adept at looking at data and making inferences about how things could and could not have occurred. The more a person studies various kinds of life, the more he or she begins to see patterns emerging and imagines what might be essential. Chapters 4 and 19 explore the current theories.

The question of life's origins touches on matters of religious import to a few readers. As a Christian, I think of astrobiology as a way to better understand how God created the world. Some scientists who are atheists think of astrobiology as a way to understand how life could come into being without a creator. I do not believe that the available data really allow us to draw a conclusion either way. With no available examples of life springing from nonlife, I think we can have very little confidence about what might or might not have been necessary. In this book I will limit myself to the areas that offer more concrete answers.

Putting the Pieces Together

Many people wonder how astrobiology can be a science when it deals with such an amorphous topic—life in space—and such a broad range of disciplines. How do all the different pieces come together, and why do we care?

Scientists in general try to understand things. It's not just curiosity, though that is a vital component. Neither is it simply a matter of knowing how to do things. Often research has the aim of producing specific ends, such as how to build a better bridge, or predict the weather, or blow something up. Other research, however, aims at nothing more specific than understanding. Humans, both as individuals and as societies, want to know how the universe works. We look at the world with wonder and ask, "How?" Imagine walking into a giant factory with strange moving parts and saying, "What is this?"

Humans have always been curious about the living world. Our lives depend on being able to tell the difference between a cabbage and a stone—is it edible, or not? Or look at the issue from another angle. Our lives depend on identifying things that make those same decisions about us, such as saber-toothed tigers. Finally, we depend more than we might think on our ability to recognize other humans—for family life, community, and any activity that requires collaboration, from making bread to making spaceships. Identifying life means more to us than an academic exercise—it means existence.

All of us, scientists in particular, try to make science sound more glorious, but I'm not sure that's necessary. We explore, and though the tests and the tools we employ get bigger and more complicated, we are still using them to taste and touch the world around us. Sometimes our desire to understand results in immediate benefits (better clothing, houses, and medicines). Sometimes it results in new powers (explosives, transportation, and communication). Sometimes the results come immediately (like TNT), and sometimes they are not appreciated fully for a century or more (like genetics).

The question—or perhaps I should say the questions—of life will not be answered easily. Humans have tried for millennia to understand how we came to be, how we fit in, and where we are going. Astrobiology's popularity stems from the fact that it promises to touch on these questions in a scientific way. Controversy arises for the same reason. Who are we to deal with questions religion and philosophy (and science) have been tackling for generations? Here we must be extremely careful to wield science appropriately. In some cases, such as the origin of life from nonlife, we may never be able to reach a scientific answer. (You never know until you try, though.) In other cases, such as the formation of stars, science has provided incredible insights.

A Plan

The first five chapters of this book address the philosophical foundations of astrobiology. Most astrobiologists recognize that we are only now laying the groundwork for a future science that will connect the realm of stars to the realm of planets, the realm of chemistry to the realm of living things. A number of researchers have begun to build bridges by collating, comparing, and connecting their results into a single narrative of life, not just as a feature of our planet, but as a function of the universe. This means we need to reflect on how our science works and what fundamental assumptions we make as we do it. The rest of the book proceeds from the general to the specific.

At the most general level, we can look at the beginning of time—how matter and energy came to be and the rules that apply everywhere. Cosmology and astronomy address these issues. The origin of the universe seems to be removed from the origin of life as we know it, yet the formation of stars and planets from the interstellar dust establishes the outermost boundary for the study of life. Thus Chapters 7 through 11 concern the research of astronomers who study the origins of stars and planets, both locally and abroad. Astronomers begin the story of astrobiology by setting the stage—little balls of rock to which life clings and giant balls of burning hydrogen to give life energy.

Next come the planetary scientists and geologists, who ask how planets operate. Atoms like carbon and oxygen have to be in the right places at the right times to interact in a way conducive to life. Complex cycles driven by planetary and solar energy churn the atoms and chemical compounds, moving all the necessary pieces into place so that life can take advantage of them. In Chapters 6 and 12 through 14, I place individual organisms in the context of time and place, relating the general principles to concrete examples in history. If we think of the planet as a theater, then geology and planetary science present the players. No action—at least no "live" action —has occurred, but the stage is set and the actors have taken their places.

In Chapter 15, I deal with questions of chemistry and the chemical foundations of life. Somehow, when we were looking the other way, the opening dance began. The players interact, but the plot has yet to become clear. Because we still have trouble defining life, no one can say exactly where abiologic chemistry ends and biochemistry begins. Suddenly we find ourselves caught up in a drama we did not expect.

Biochemists and biologists add the decisive plot elements. Life, once begun, acts in surprising ways. Simple chemistry becomes more complex. Organisms learn to trap more and more solar energy in their systems. They grow and divide and multiply. In Chapter 16, I discuss how the astonishing amount of available energy—provided by the Sun and Earth— gets incorporated into the biosphere. The plot becomes more interesting. What seemed to be a single storyline has diverged and developed until we find it difficult to keep track of all the separate plotlines. Birth, death, invention, alliances, and betrayals all take place at an amazing speed.

Ecologists, paleontologists, and evolutionary biologists start to put things back into perspective again. Ecologists study how organisms interact in space while the latter two groups study how they interact through time. In Chapters 17 and 18, I address questions of how we can see the bigger picture through all the apparent chaos.

And, even though it seems we have left behind the earlier elements, they always come back into play. Ecosystems grow large and change the chemical composition of atmosphere and crust, affecting how much and what kind of energy can enter the planet. Organisms produce byproducts that change the chemistry of land, sea, and air. Perhaps life will one day even break the confines of the theater, leaving Earth. All of these interactions provide opportunities for astronomers, biologists, and a host of other scientists to compare notes and make new discoveries.

The last two chapters return to philosophy and religion. Having dealt with what science can say, I will reflect on how that might impact the way we see ourselves, for humans act both as players and as audience. Of course, this takes us beyond astrobiology proper, but any good scientist, having reported the research and clearly conveyed what is known, asks what comes next and how we might approach it in a meaningful way.

Many people study astrobiology in some form or another—I know of at least 1,386. They have their own perspectives and insights. This book tells you what makes the subject fun and compelling to me. If I have caught your interest, please read some of the wonderful publications produced by astrobiologists around the world.[2] Many great resources are available online.

It has been my experience that learning astrobiology can be extremely difficult because there are so many interrelated systems to be considered at once. Big-picture questions like evolution and ecology only become clear when standing atop a mountain of data. College-level biology students often complain that the material seems to involve too much memorization. Even the ingrained habits of different scientific disciplines may clash; for example, physicists often accuse biologists of stamp collecting (putting together long lists of anecdotes rather than running on iron-clad laws). I can only say that addressing the big picture and the minute data at the same time seems to be the best way. We want to summarize the grand scheme of life in space, but we have to resort to a huge number of particulars.

This book moves sequentially from the oldest and biggest events to the most novel and particular organisms. Such a scheme still could not encompass everything I felt essential, and so there will be some disorder in my story. Most notably, I must begin with several philosophical and practical questions about the exploration of space so that, arriving finally at real organisms in Chapter 9 and more uniquely human concerns in Chapter 20, you may see them in a new light.

2

Living Science

Scientists and nonscientists often fall into the trap of believing that all science proceeds in the same way. Yet a number of very clever people conduct extraordinary research in a variety of ways. One of the biggest challenges in developing astrobiology as a field has been to get all the clever people to speak the same language long enough to make progress.[1] Nevertheless, most scientific inquiry proceeds from a certain set of basic assumptions.

What image comes to mind when you think about science? If it is a serious man in a white coat and glasses writing equations on a chalkboard, you are not alone. This was a popular image in the twentieth century and has entered into the public consciousness, both as icon and joke. Scientists, of course, can be a great deal more colorful, and we've come to expect them to argue with each other, both in scientific papers and on the nightly news. As time passes, more and different people enter into scientific research, doing things in increasingly diverse ways. Ah, you say, but the basic methodology remains the same. And you would be right. Science rests on some important basic principles. First, science involves discussion. Second, scientists make predictions. Third, the predictions fail more often than not. Fourth, and perhaps most important, science presumes that the universe behaves itself.

So what does all that mean?

First, science involves discussion. Scientists talk about things other scientists can perceive. Contrary to popular belief, science does not say that if only one person sees something, it must be an illusion. Rather, scientists know better than to discuss something unless we know that enough people see it to make discussion worthwhile. I like to think of it as discussion about "mutual observables." Beyond the issue of whether a discussion is worthwhile, science rests on the assumption that two people are less likely than one to fall into the same trap, whether based on flawed perception or faulty reasoning. In fact, the more people who consider an idea, the better. It doesn't mean that the group always gets it right; it is simply that a group gets it right more often than an individual. Different people bring different perspectives, more data, and more information.

Albert Einstein said, "Common sense is the collection of prejudices acquired by age eighteen." Common sense is not a bad thing. It represents collected wisdom distilled through our memory and reason. It is, however, highly dependent upon circumstances—whatever has been common in your life. Reason allows us to reflect on our assumptions and see if they are consistent. Observation allows us to test a belief against reality. Scientists bring our common sense, reason, and observation together to assess whether one idea is better than another.

Second, science involves making predictions. Predictions sound like this: If you sail off to the east, you will eventually come back around the world from the other side. The value of such statements in science comes from how often they accurately predict or explain events. Some fields of science are built around laws that make definitive, exclusive predictions: Planets always travel in elliptical orbits around their parent star. Objects in motion will always stay in motion (except when acted on by an outside force). Certain predictions work like laws—they operate everywhere all the time.[2] If we see even one event contrary to a law, then the law gets tossed out. Scientists also make statements called probabilistic laws. They predict how a large number of events will occur over a certain period of

LIFE IN SPACE

time. Every 5,730 years half of all the carbon-14 atoms in any group will turn into nitrogen-14.[3] It cannot be said which of those atoms will decay, only that, in a given period, half of them will. Scientists can also identify some general tendencies—things that seem to occur frequently, but the details of which are not understood. We know, for example, that plants do not eat insects—except that in rare instances they do. It is possible to divide our laws into a number of categories: definitive or absolute laws, probabilistic laws, and predispositions.

Physics benefits from the study of countless very small, very simple objects and countless very brief, very common events. For this reason, it possesses quite a few definitive laws. After a billion or so observations of a phenomenon, you can be fairly confident that you have a good grasp of how it works. Biology, on the other hand, deals with some very complex systems that almost invariably fail when separated. Fewer completely independent observations lead to lower certainty and fewer universal rules. This is not to say that biology is not as well developed as physics, only that the questions are different and less amenable to the type of exhaustive, controlled analysis that has proven so successful in that field. We can see that physics and biology operate differently. Both deal in common perceptions and general principles, but one has far more definitive rules than the other.

Third, predictions fail more often than not. This has proven one of the most difficult aspects of science for nonscientists to understand. Every good prediction follows on the heels of at least a few (sometimes hundreds) of not-so-good predictions. In fact, the best predictions often result from the endless modification of earlier theories that had to be revised over and over as new data became available. Gravitation and evolution both exemplify theories valued because their predictions are so much better than those offered by any alternative theory. In both cases, many other theories failed before the current best one came along. Both will probably be replaced by better and more accurate theories later on, but don't expect this to happen anytime soon. After centuries of observation, these theories explain the data exceptionally well, and, to date, have been able to account for every observation.

The process remains crucial. Science is the process of making predictions, matching them up to observations, and seeing what happens. Most of the time the predictions don't match very well and new predictions need to be made. Much less often, they don't match and new observations need to be made. Rarest of all, they match up perfectly and everyone gets to have a glass of champagne and go home for the day. But for the most part, the path toward a deeper understanding is long and strewn with discarded predictions.

Fourth, science presumes that the universe behaves itself. One of the central yet often ignored assumptions of science is that the universe is roughly the same here and there, now and then. If I make observations today and tomorrow or on Earth and the Moon, they will turn out the same. We call this assumption the principle of symmetry. It means that, all other things being equal, one place or time is interchangeable with another. Symmetry breaks down on a rare occasion. Chapter 4 takes a deeper look at when and where symmetry applies, and why astrobiologists need to use it cautiously. We are beginning to appreciate that the basic rules are different in bizarre instances like quantum singularities. Likewise, "life as we know it" need not be "life as it can be known." Still, for the most part, the universe behaves consistently. When I observe something and you observe something, we have a similar enough experience to talk about it and make predictions. So we come full circle. Science involves groups of people sharing observations, making predictions, checking the predictions, and then making new and better predictions—all the while living in a universe that tolerates this sort of behavior.

Experiments

I have been exceedingly careful not to use the word experiment in my definition of science. You may be wondering about that. Aren't experiments essential to science? Not necessarily. Science can be divided into two rough categories, experimental and historical. Both entail observations and predictions, but do so in slightly different ways.

Experimental science usually happens in a laboratory, although it can happen anywhere. In this type of science, an activity is repeated a number of times, often changing a single aspect of the activity to see whether events occur in the same way each time. The classic (if apocryphal) example has Galileo dropping two weights from a tower to see if the heavier one hits the ground first. Physics professors frequently recreate this event for their first-year classes with a feather and a lead weight dropped in vacuum tubes. Acceleration due to gravity is roughly 9.8 meters per second per second on Earth, regardless of the size, shape, or weight of the object dropped. All other things being equal, two objects will always fall at the same speed. This is a great experiment, because you can change one thing at a time—whether it be weight or mass or shape—and all objects will fall at the same rate.[4]

A great deal of science is conducted under these conditions. If a number of causes seem to be linked to one effect, the causes can be removed one at a time to see which one (or more) of them is unnecessary. One of the great benefits of space-based research (on the Space Shuttle or a space station), is the ability to change the force of gravity in a way that is not possible on Earth's surface. (Space-based experiments related to astrobiology have demonstrated how plants sense gravity.) Physicists and chemists regularly do experimental science.

On the other hand, a great many scientists engage in historical science. They look at events that have already occurred—hence historical—and try to explain them. Though this may sound unscientific at first, it is both common and methodologically rigorous. Often one observation can be used to explain another. Scientists may try to generate theories that allow them to predict one measurement, such as the mass of a star, based on another measurement, the wavelengths of light that the star emits. They predict the observation, but not the event. After all, the star's mass will have been determined years ago.

The most well-known examples of historical science come from astronomy. Celestial objects are too far away to manipulate for experiments. In many cases, even if we could reach them, the trip would take so long that we would not be able to use the results anyway. It takes light a little over eight seconds to travel to Earth from the Sun, and over four years from the next nearest star (Proxima Centauri); therefore, everything we know about these stars happened in the past. Likewise, we know a great deal about the other planets in our Solar System based on information gathered from Earth. In the seventeenth century, Johannes Kepler used observations of local planets to predict how objects orbit the Sun. All other objects observed since then—extrasolar planets as well as countless asteroids and comets closer to home—follow the same laws. This is a great example of using one observation to predict another, even though all the events happened in the past.

Reductionism

Another common division among sciences—and some scientists—involves reductionism. Many scientists believe that all complex processes can be explained based on less complex processes operating at a lower level. Humans are fueled by eating plant and animal matter (biology). On a smaller scale, this can be explained by the ingestion of carbohydrates. Cells in the lining of the digestive system absorb the carbohydrates and break them down for energy (histology and physiology). On a smaller scale, one might say that carbohydrates break up into glucose molecules and glucose is degraded, releasing adenosine triphosphate (ATP) and nicotinamide adenine dinucleotide (NADH) to the cell (biochemistry). On a smaller scale, electrons are transferred from ring orbitals in the glucose to lower-energy orbitals in the bonds connecting phosphorus, oxygen, and hydrogen atoms (chemistry). On a smaller scale, it is even possible to look at how the electromagnetic force binds electrons to nuclei in all of the molecules (physics). In each case the bigger phenomena can be explained and "reduced" to interactions among smaller, more common players; hence reductionism.

Reductionism has been extremely influential in science and has dominated the field for around one and a half centuries. Better understanding of smaller and smaller phenomena has led to increased understanding of larger phenomena. We see the impact of subatomic particle interactions on atoms, the impact of atoms on molecules, the impact of molecules on chemical systems, and on and on all the way up to the impact of organisms on ecosystems and the impact of ecosystems on the planet. The greatest benefit comes from the fact that we can often eliminate thinking about complex processes and focus our attention one level down. Great progress was made in biology once it was realized that heredity (the passing of traits from parents to offspring) could be explained by molecules called genes, and no "life essence" was necessary.[5] Great progress was made in chemistry once it was realized that different elements, such as carbon and nitrogen, differ fundamentally in the number of protons they have in their nuclei.[6] Both of these highly influential discoveries occurred gradually throughout the late nineteenth and early twentieth centuries. Reductionism works.

Reductionism faces two major problems, however, the first ideological and the second practical. First, how far down can you reduce? In practice everything gets reduced to the interactions of subatomic particles. The particle physicists are thrilled about this, as it adds a great deal of prestige to their field. Why should it stop there, though? If string theory ever becomes a practical way to explain particle interactions, then the string theorists will be the most prestigious.

Why do we favor simpler interactions? They are easier to study—better behaved. It is also comforting to believe that the universe works like a machine, based on smaller and smaller cogs. We don't know where it ends, however, and one begins to suspect that knowledge at different levels may be useful for other reasons.

Second, many complex phenomena simply cannot be explained at a simpler level. Biology can make a large number of useful predictions about the composition and behavior of animals that do not reduce. Psychology, anthropology, and economics tell us things about human behavior that cannot be reduced to biology. I have a personal fondness for sociobiology, which attempts to reduce behavior to evolutionary interactions, as long as it is tempered by the humility that says better predictions can still be made at a higher level. Why would you want a bad molecular theory of behavior in place of a good sociological model? Reductionism constitutes, at the very least, a desire to explain events at the smallest level possible. It need not represent a disavowal of higher-level explanations.

On the opposite side of the coin lies emergence theory. Believers in emergence say that some properties of complex systems result from synergy—the whole exceeds the sum of the parts. Scientists debate what emergence means. It could be the recognition that some things are currently irreducible (what I'll call weak emergence) and should be studied predominantly at the higher level, for the moment at least. It can also be a statement that some things cannot be built from the bottom up (strong emergence). Intelligent design arguments—life and sentience are so complex that they could only be generated from pre-existing life and intelligence—use the notion of strong emergence as a central tenet.

Completely weak emergence provides limited utility. If we simply put off our reduction to some point in the future, we still have reductionism as the ultimate and best goal. On the other hand, over-strong emergence theory makes it too easy to pick a phenomenon and say it is beyond explanation. That has no utility at all. Science must attempt to explain. By my definition, science makes predictions. That should be the goal, even if it seems impossible to achieve.

Some scientists will probably deride me for considering emergence at all. It is exceptionally unpopular in some circles. The idea cannot be avoided in a book on astrobiology, however, for the simple reason that we currently cannot generate life from nonlife. I do not say that it will never be done; several labs are making interesting discoveries. Still, we don't know how to do it, and the problem is quite complex. Although reductionism has taken over in most of the natural sciences, biology is stuck with the consistent observation, best made by Louis Pasteur in 1859, that spontaneous generation—life from nonlife—does not occur. A large number of scientific predictions regarding health, food preparation, and medicine depend critically on this fact. If you can get from nonlife to life, it must be extremely difficult. Biologists will not be able to completely embrace reductionism until we resolve this problem. Multicellular organisms and quorum sensing—when large numbers of unicellular organisms behave in concert—may also provide fruitful ground for emergence research.

At the same time, I want to state very strongly that emergence should never be used as an excuse to stop looking at a problem. Emergence will never be a successful argument for dismissing evolution. We cannot explain how the very first organism on Earth came to be, but once it exists, evolution explains the rest quite satisfactorily. Modern theories of evolution make incredibly useful and consistently accurate predictions, and doubt about reductionism in no way diminishes this.

Emergence theory provides an opportunity to look at new avenues of research. Information theory gives interesting insights into how systems behave once they reach a certain level of complexity. The rise of deoxyribonucleic acid (DNA), for instance, provided a whole new language for biological expression. Information can be conveyed more succinctly, and life requires less energy to replicate itself. Emergence seems to be a useful way of thinking about energy, entropy, and information storage in complex systems.

In short, reductionism and emergence both have some utility in a discussion of astrobiology. Although I lean heavily toward the former, it does not adequately address every problem. The extent to which scientists rely on reductionism varies from field to field.

Hard Science

It has become popular to refer to physics and chemistry as "hard sciences" because they deal with such well-behaved phenomena and because they tend toward experiment and reductionism so well. I have also heard the terms "exact" or "true" science recently. Certainly it is extremely important to be aware of how we go about research, but it can be misleading to think that scientists in these fields have a better understanding of their subject matter than their colleagues in less reducible fields.

All science involves the scientific method, observation, discussion, predictions, changing your mind, and a fundamental commitment to an understandable universe. Nonetheless, standards and specific methodologies differ. As astrobiologists from different fields talk to each other, they experience culture shock. A physicist might not consider a sociology experiment to be as rigorous as it should be, whereas a sociologist may not think a physics experiment particularly relevant or interesting.

Here I must introduce the concepts of "epistemology" and "confidence." A branch of philosophy, epistemology deals with how we know what we know—the very careful examination of how people, including scientists, come to conclusions. A related concept, confidence expresses the extent to which people are willing to rely on a statement. Remember the discussion about scientists making incorrect predictions and changing their minds? Confidence has everything to do with changing your mind. If I have a prediction supported by 2 million observations that contradicts a new observation, then the first thing I would do is recheck the observation. I place a high level of confidence in those theories and predictions that explain large numbers of observations. On the other hand, if a prediction is new—perhaps it was created six months ago, when I thought it was a particularly clever way to explain something—I would be far more likely to question it right away. Science involves a process of prediction and observation that leads to increasing confidence. Whatever theory currently leads the pack, it should have greater confidence than previous theories simply because it was created in light of more observations. Confidence tells us how good we think a theory is. Epistemology studies how we generate confidence.

Moving back a few steps, evolution involves three important theories— "heredity," or the transmission of information to offspring; "variation," or the diversity of offspring that results from imperfect transmission of information; and "selection," or the failure of some organisms to survive and reproduce. Each of these theories makes strong predictions that have been correct uniformly and exclusively for all recorded observations. That constitutes a huge body of data and, consequently, very high confidence. If we say, "evolution is just a theory," we must at the same time admit that it is an exceptionally good theory. It explains a vast number of observations.

On the flip side, astrobiology has to deal with questions like this one: Is there life on Mars? In 1996 a team of scientists announced that they had found evidence of martian life based on the martian meteorite ALH84001.[7] Indeed they had—found evidence. A number of observations were consistent with life on Mars. Life on Mars, however, is a truly extraordinary claim. We find the idea monumental. David Hume once suggested, "extraordinary claims require extraordinary proofs." Thus confidence in the proposition for life on Mars will come at a high cost. In the end, the scientific community decided that ALH84001 was not enough. It was a perfect example of the scientific method. Observations were made. Explanations were constructed to match the observations. Scientists talked to each other. Many people admitted they were wrong about things they had said in the past. Predictions changed. Most important, scientists discovered a whole new suite of possible questions to ask and ways to investigate life.

Several features of ALH84001 would have been considered evidence for life, if the meteorite had come from Earth. Because life on Mars was such a significant claim, however, physicists, chemists, biologists, and geologists devoted a great deal of effort to looking at whether using those observations as evidence for life was a good idea. Some specific theories about life had never been tested for such a foreign environment. In the end, the consensus (it was a discussion, after all) formed around the idea that the meteorite contained evidence for life on Mars, but not sufficient evidence for any level of confidence worth mentioning. It was a starting point.

ALH84001 is an important foundation stone for the study of astrobiology; it forced us to ask several questions. What would it mean if life existed outside Earth? What would it take to convince us that it did exist? How would such a statement interact with all the other things we think we know about the world? These questions must be considered by those in a number of scientific disciplines. The challenge will be to integrate them in a meaningful way, so that we can pull puzzle pieces from each and put together a "big picture" theory of life.

3

Defining Life

What makes a thing alive? When you encounter something new, how do you tell? The question is surprisingly difficult. Start with an example so old and so common it might not have occurred to you to think of it. Are eggs alive? The straightforward answer is yes . . . and no. Let us stick with chicken eggs for the time being. I would be willing to bet that almost all of my readers have encountered chicken eggs and the majority of you have eaten them. Are chicken eggs alive?

One immediate answer is yes. Of course eggs are alive. Where do new chickens come from, after all? Setting aside the question of which came first, chickens produce eggs, eggs hatch into chickens, and the cycle goes on and on. Eggs function like little incubation pods where developing organisms grow. In the case of chickens, fertilization occurs inside the chicken, and a new little *Gallus gallus* forms. It starts to develop inside a protective shell produced by the mother. There was a time when that may have been a sufficient answer. Once, all eggs were fertilized eggs, chickens in the process of becoming. In the United States and many other countries, however, we mostly eat unfertilized eggs. Large battery farms keep the male and female chickens separate and the females continue to produce eggs. In this case, the egg is nothing more than a really big cell with some support tissue and a hard coating.

If this concept seems strange to you, you are not alone. Even as a trained biologist, I have to wonder a little at a species that puts that much energy into a cell that won't contribute to the well being of the mother or any offspring. Chickens have been bred by humans to produce large, unfertilized eggs on a regular basis. We like them that way. Chickens are not alone. They produce unfertilized eggs to feed humans, but other species produce eggs to feed their own offspring. Tree frogs are a good example. Some species use unfertilized eggs much as mammals use milk. So the question remains. Are eggs alive? They can be. In some cases they are a vital link between generations of chickens. In other cases, they are simply waste products, like human skin cells that fall off and are no longer alive.

The problem rapidly becomes quite complex. How do we decide what constitutes an individual organism? Biologists themselves continue to debate definitions, not only for life but also for organism and species, even individual. Even for a large animal like a chicken—well characterized and common—we still do not have clear answers.

This chapter deals with what it means to be alive. (The question of individuality must be saved for Chapter 14.) I begin by writing about some of the more common—and useful—definitions of life that have been popular in recent years. I discuss advantages and disadvantages of each. To date nothing truly satisfactory has been proposed. Afterward, I provide some guidelines about how I think the problem should be addressed and discuss the progress that has been made by those in astrobiology and related fields.

Astrobiologists require a better definition of life than any of those currently available. What we want is something diagnostic, a way to make predictions about living things and a measure of "liveness." In the absence of such a tool, however, astrobiologists can say a great deal about the story of life in space. People have been peeking at, prying into, and trying to wrap their heads around the problem for a long time, and some interesting things have been discovered. The remainder of the book explores the question "What is life?" not by categorization, but by example.

The Pornography Definition

Perhaps the most common definition of life has come to be known as the pornography definition. Although life on Earth does involve sex, this definition comes from an entirely different trade, the field of law.

The crux of the pornography definition involves a statement made by Supreme Court Justice Potter Stewart: "I know it when I see it."[1] This phrase makes an appeal to common sense and can be extremely useful. I might say when cleaning out the refrigerator, "Ack! The coleslaw is alive. I'm throwing it out." To which you would be perfectly happy to concede the common-sense definition. I know life when I see it. Likewise, having picked up an object on the beach, I could fairly confidently determine if it was alive. Figuring out how to tell the difference between living things and nonliving things—preferably before you try to eat them, or vice versa—can be considered a survival skill. Bearing this in mind, the pornography definition can be both useful and necessary in daily life.

On the other hand, scientists tend to be extremely skeptical of common-sense appeals. Common sense is not common to everyone, and sometimes it's wrong. For example, I can think of a few things that should have been thrown out of the refrigerator, but were not. You might also recall the case of ALH84001, the Mars meteorite. Common sense fails because extraterrestrial rocks are uncommon. We can expect extraterrestrial life to be even more so.

William Brennan, a later member of the high court, expressed the fundamental problem with the pornography definition: "none of the available formulas . . . can reduce the vagueness to a tolerable level."[2] Pornography definitions tend to be broad, vague, and open to debate.

Astrobiologists, like many scientists, lean away from the pornography definition. If science involves discussion, then we must learn to use a common vocabulary. Although "I know it when I see it" works for individuals, groups have difficulty agreeing on a common meaning. Two incidents have proven that the pornography definition leaves much to be desired.

The first case, discussed above, relates to the question of life on Mars. Does the meteorite ALH84001 provide evidence for life on Mars? Consensus says probably not, though some scientists continue to investigate the issue, hoping that future findings will be more definitive.[3] The second case involves the earliest evidence for terran life. When looking at remnants of life from billions of years ago, how sure can you be that you have interpreted the data in the best way possible? Stephen Mojzsis claims to have found chemical evidence for life prior to 3.8 billion years ago.[4] William Schopf claims to have found fossils that formed 3.5 billion years ago.[5] In both cases, serious critiques continue within the scientific community.[6]

So, we see that the pornography definition can be really useful when we're dealing with common things. Life found on the beach or in the refrigerator matches the definitions of life we have been slowly putting together in our heads all of our lives. The pornography definition fails, however, when we come to the edges of our experience, which is precisely where astrobiologists want to be. When looking at the earliest or most obscure or most alien forms of life, our common sense cannot help us. It has not had sufficient practice. In looking for a definition of life in space, it will be necessary to turn elsewhere.

The Biochemistry Definition

Another common definition has to do with biochemistry. Life as we know it involves a number of apparently unique chemical pathways. Terran organisms have a real talent for stringing together carbon molecules and making new organisms and materials, including such complex and naturally improbable compounds as wax and honey, enamel and milk. Biochemical similarities can be exceptionally useful for exploring the history of life on the planet, and may be applicable elsewhere in space. As we will see in Chapter 6, some chemical principles may be universal to all life. As far as we know, there are a limited number of ways to capture and use energy in any manner recognizable to humans.

LIFE IN SPACE

A biochemical definition of life seems promising, but once again we run into the problem of familiarity. We know that certain terrean organisms use certain biochemical pathways, but that does not mean that life elsewhere in the universe operates in the same way. Even if we could be sure that it did, the biochemistry definition would still fail us. Before the nineteenth century, many scientists believed that living chemistry was fundamentally different from nonliving chemistry. The term "organic chemistry" was coined to note this difference. More recent discoveries have shown that no clear dividing line exists. Organic chemistry has come to mean carbon-based chemistry.[7] Thus, biochemistry (the chemistry of life) and organic chemistry (the chemistry of carbon) should not be confused.

To the best of our knowledge, biological reactions in chemistry do not differ from abiological reactions in any sense other than speed and probability. The sentence contains more information than may be immediately obvious, so let us unpack it. Fundamentally, biological reactions follow the general rules of thermodynamics. Thermodynamics involves all scientific explorations of how heat interacts with matter and other forms of energy. Chemists use thermodynamics to figure out which chemical reactions will occur. Some reactions occur "naturally" when two molecules come together. If you have ever mixed vinegar and baking soda, you have seen how they recombine and release carbon dioxide bubbles. On the other hand, some reactions need to be encouraged by adding heat. A good example of this can be found in clay pots. Clay includes compounds of aluminum, silicon, and oxygen, with some water added. When a potter fires molded clay, the water evaporates and the heat gets absorbed in bonds created between aluminum and silicon molecules. This produces a new, stronger structure as the baked clay hardens. Thermodynamics predicts which reactions happen on their own and which require heat based on the energy trapped within different compounds and released by different reactions.

Life does not, indeed cannot, make thermodynamically impossible reactions occur. It can facilitate processes by applying order and heat to make reactions occur more quickly, more predictably, and more often. Life invests energy to see that things happen in a fixed and definite way, organizing molecules and making them more likely to react.

Unfortunately, this facility of life for speeding up reactions makes it difficult to differentiate between life and nonlife without a stopwatch. When looking at possible evidence for life, scientists need to know if a compound formed within a matter of seconds (life) or a matter of centuries (nonlife). The debates over early evidence of life on Earth revolve around these very issues, as researchers attempt to reconstruct the environment in which fossils formed. The answers seldom come easily.

Likewise, life tends to concentrate molecules that occur only rarely in abiological conditions. Many forms of rock on Earth—shales and limestone, for instance—have very dense concentrations of carbon compounds. It seems unlikely that such large quantities would be concentrated in one place without the presence of past life. On the other hand, over the past centuries, geologists have come to understand that certain millennial processes form carbonaceous chondrites—meteorites with concentrations of organics that only form after thousands of years in space. We know that the abiological process occurs, but we also know that the biological process happens faster.

All of this makes it extremely challenging to separate biochemistry from all other forms of chemistry. An individual reaction may be the product of biological action or simply a random (if rare) abiological process. We can only tell the difference by looking at reactions in context. How often do they occur, and how quickly? In the twenty-first century, chemistry continues to become more and more sophisticated. I have little doubt that definitions of life will include chemical elements, but there is currently nothing unique about *bio*chemistry that would allow us to use it as a definition of life.

The Anti-entropic Definition

Thermodynamics does suggest one possible out—entropy. Entropy can be thought of as the opposite of order. Imagine a deck of cards organized by

LIFE IN SPACE

number and suit. Every card has a place; the whole stack is neatly piled in your hand. Now toss the deck up. As the cards tumble through the air, they lose their order. When they finally reach the floor, they have little or no order left. That is entropy, and thermodynamics tells us that (for a closed system) disorder always increases. Looking at the entire universe, order disappears constantly. If we want to build something up—if we want to increase order locally—we must invest energy or heat. That energy must come from somewhere else—somewhere that loses order.[8]

Being alive and being surrounded by living things, the concept of entropy may seem counterintuitive. We organize the world. It is something organisms do. Every time you eat, however, you break down something in order to build yourself up. We draw energy from one reservoir and use it someplace else. Organisms exist in perpetual balance between their ability to acquire energy and their need to spend it in self-maintenance.

The Sun burns hydrogen and helium, releasing tremendous amounts of energy in the form of waves—light and heat. Light hits Earth, where bacteria and other primary producers convert it into chemical energy. The bacteria form carbohydrates and other complex biomolecules. We consume the carbohydrates (or protein from animals that consume the carbohydrates) to fuel our own systems. Entropy pulls downward, but, by the constant efforts of life, order is maintained. Order increases for individuals all the way up the ladder, but entropy increases for the whole system (that is the universe) throughout, because energy dissipates at every stage. It takes more energy to make sugar than you get from eating it.

Coming back to the definition of life, some astrobiologists have recognized that the terran biosphere involves tremendous amounts of order. Life puts things together to make patterns and then maintains the patterns over time. Arguably, an alien arriving at Earth from any distance would know there was life here. The order is so incredibly high it would be hard to miss.

Is it possible that we could look for anti-entropic systems on other planets? Yes and no. Low entropy would be a great sign of life and should be investigated, but it may not be conclusive. The problem lies in measurement. At a very large scale, entropy always wins, but at what scale do we look for anti-entropy? Crystals, for example, have a high level of order and can form slowly over time. This would seem to be a case of local decrease in entropy. And yet, we do not consider crystals to be alive. Likewise, stars show a great deal of local order, but we would never call them biological. High levels of order (low levels of entropy) may be suggestive of life, but cannot be definitive.

The Replication Definition

Certain types of order show more promise. No one debates that life is capable of replication. All known types of organisms reproduce themselves. Humans make more humans, bees make more bees, and bacteria make more bacteria. Even the more exotic entities such as viruses and prions—which may or may not be alive, depending upon our definition—replicate. So, whatever it is, if it makes more of itself, it is alive, right?

Alas, no. The first critique relates to the nature of replication. Scientists have noted that clay minerals and crystals are capable of expanding by building on an existing template. In many cases, environmental conditions mean that the growing structure will fragment and the pieces will continue to grow. The process looks remarkably like biological replication.

The common-sense response to the first critique is this: "Yes, it looks similar, but I can tell the difference. After all, crystals replicate identically and with very little structure, whereas living things have great variety." Like all common-sense definitions, it works well on common objects of study. Almost no one would mistake a bee or a worm for a crystal; their replication processes are markedly different. Even bacteria usually look much different under a microscope.[9] The question could be debated, but let us assume for the moment that a knowledgeable observer can tell the difference.

Another problem remains, and I consider it far more important than the first critique. How do you tell if you are looking at the right sample? A donkey can reproduce, and so can a horse, but when you cross them the offspring are sterile. Would you consider a mule alive even though it cannot replicate itself? Overall, reproduction for a species seems obvious, but it can be tricky for individuals. Take another example, the bee. Most bees in a hive can reproduce, but do not. Only the queen and drones reproduce. Are the workers alive? And the trickiest example yet: multicellular organisms. In humans, germ cells (in the reproductive organs) replicate and contribute to the next generation of humans, but most of the cells in the body do not. Does that mean that humans are partially alive? Are some of our cells more alive than others? If we limit life to replicators, how can we judge what level a new object represents—cell, organism, family, or community? We still cannot say definitively if it is alive or not.

Replication may not provide the answers we want. All types of organisms reproduce, but not all individual organisms. When looking for life in space, which is more likely—to find a whole type of organism, or to find an individual? Some theorists have even speculated on the possibility of a single living entity that persists through time, changing but never reproducing. If the population of a species is one, and the individual lives for centuries, should you call it alive?

What about self-replicating machines? How would we categorize robots that make more robots? This seems to be a standard science-fiction topic. Fortunately, astrobiologists have a simple response to this question—we avoid it. We are searching for evidence of life. Self-replicating machines, whether or not they are alive in and of themselves, must have been constructed by living things. They must be evidence of life. So the astrobiologist's answer is that self-replicating machines are good enough.

The Evolutionary Definition

If perfect replication is out, perhaps imperfect replication might not be. Evolution seems to be a common property of all life. Evolution, as I stated in Chapter 2, has three central components: heredity, variation, and selection. Since selection seems to be a property of the universe, rather than organisms, the first two elements may be used as an evolutionary definition of life—heredity and variation. If I were to speak in strictly Darwinian terms, I might say "descent with variation." The definition requires living things to imperfectly replicate themselves.

There seems to be a fine balance between perfect replication and random change. Crystals can be said to replicate perfectly, but they cannot evolve, because they only conform to a limited number of states. When errors are introduced, the crystal stops growing or returns to the original pattern. Organisms, on the other hand, gradually change over time, resulting in abundant variety. It must be noted that entropy prevents the perfect transmission of data. No matter how much energy you put into maintaining a signal, errors will creep in. In information theory we solve the problem by transmitting multiple copies of the same message so that errors can be found and removed. Many organisms use the same technique, keeping several copies of their genetic information handy to prevent mishaps.[10]

Perfect replication, of course, is both costly and less than ideal. Why would an organism expend massive amounts of energy to produce a perfect clone when it may be preferable to produce two (or two thousand) slightly different offspring that may be better at surviving? Many organisms have developed complex mechanisms whose sole purpose is to jumble the DNA of their progeny. Sexual reproduction, quite familiar to humans and other mammals, involves the random recombination of maternal and paternal genes. Some bacteria swap DNA packets using viruslike proteins or building bridges between cells.[11] Organisms embrace variation because it can speed up the process of change to fit an environment. Don't keep all of your eggs in one basket and don't keep the same genes in all of your eggs.

Although evolution can be a difficult process to define, it clearly follows a pattern of imperfect transmission resulting in changes with time. Anything that does this could be said to be alive. I have a certain fondness for

LIFE IN SPACE

this idea. It is concrete. It is observable. It both matches all life as we know it and presents clear predictions for judging life in the future. On the other hand, it suffers from the same problems of scale found in the biochemical definition and the anti-entropic definition. How do we know whether or not we have a big enough sample to observe evolution? How long does it take? If we found a sample of alien matter that was alive, would we be able to watch it evolve? What if it was dead? What if it spent 100 years between reproductive cycles? What if it lacked a mate? The problems are numerous.

Even though we have found a good theory for defining life, we have failed to make a practical diagnostic. We want to be able to look into space and answer the question, "Are we alone?" We want a tool that, when we point it at something, tells us yes or no. None of the definitions proposed so far—both in this book and in the scientific discussion at large—provide that kind of certainty. We still don't have a meaningful answer to the question, "What is life?" Nonetheless, the discussion helps us to clarify what we know and believe about life.

I would like to take a closer look at two ideas introduced in Chapter 2 that I think will prove essential to our understanding. They might even form a new paradigm for life sciences. Apparently, nothing less will provide us with the tools we need to move forward.

Emergence and Reductionism

Some thinkers at the frontiers of biology and philosophy have decided to re-explore the notion of life as an emergent phenomenon. Despite conventional and historic wisdom in biology that life cannot come from non-life, researchers are starting to wonder how the whole process started in the first place. Could there be some property of energetic systems that allows them to take on the properties of life when enough energy is involved?[12]

Unless we are to claim divine intervention in some mysterious and impenetrable fashion, it will be necessary to tackle the life versus nonlife divide. That question must be tightly interwoven with how we define life. Current attempts to bridge the gap focus on elements of quasi-biotic chemistry. Could life have existed with a limited biochemical repertoire? Could it have existed using fewer kinds of molecules than it does today? These are experiments that can be carried out in the laboratory, and a number of researchers are trying to make simple systems that rest somewhere in between.[13]

This movement to build life from the ground up suggests a reductionist definition of life. Carol Cleland and Christopher Chyba advocate for a revolution in biology that parallels the revolution in chemistry that came from the periodic table.[14] Can we create a definition of life based on some new understanding of its principal parts? What new understanding will provide us with the predictive power we need and how sweeping must be the changes to our current framework of understanding?

Our dilemma is that we want life to be a binary operator—alive or not-alive—without ambiguity. At the same time, we are trying to come up with a way to understand life from nonlife. It cannot be both ways. Either life happens partially and gradually, or it springs up in some emergent fashion once a threshold has been crossed. Ultimately, the question of how life comes to be relates to whether and where life exists elsewhere. If life springs up easily, we will expect to find it everywhere. If it endures for billions of years and can survive space travel, we will expect to find it everywhere. On the other hand, if it is rare and travels poorly, we may not be able to see enough of the universe to find other examples, even if they do exist.

Personally, I find myself highly skeptical of emergence and reductionism in the definition of life. In both cases we seem to be waiting for science to redefine the terms in a meaningful way. In emergence, there seems to be a desire for the uniqueness of life to be intrinsic and irreducible. In reduction, there seems to be a desire for the problem to go away. Even though

LIFE IN SPACE

neither idea provides a fully satisfactory answer, both allow us to stretch the problem and the data, in hopes of finding a better understanding.

After millennia of thought, you would think someone would have a good and useful definition of life. We know that a number of things on Earth are alive, and we know a great deal about how they work. All share position in space, biochemistry, composition, and a unique history on a (so far) unique planet.

What can astrobiology do? A great deal, actually. We can trace the history of life on Earth backward—if not to the beginning, at least for 3 billion years. We can trace life outward—if not to the ends of the universe, at least to the edge of the atmosphere and perhaps beyond. We can ask what life does to protect itself and how far from home it can travel. If we want to know our limits, we must test them. Above all, we can look for patterns. We find ourselves caught between the despair of entropy and the hope of life. Life is interesting precisely because it is precarious, because it is hard to define. Life has a story to it, history and character. Astrobiology, for me at least, is all about discovering that story.

4

A Well-Behaved Universe

Looking at life in space, we see that our terms can have a major impact on how we ask and answer questions. Before we look for life out there we should ask what we mean by "out there." So let us review what can be known about the universe at the most fundamental level. These ideas form the basis for science in general and astrobiology in particular.

The question on the table is this: Does the universe behave itself? Does it follow its own rules? Are there consistent laws that allow us to make predictions about distant times and places? Certainly. Daily life requires some assumption of regularity for peace of mind. What happened yesterday will happen tomorrow. The way things work at home will be roughly the way they work in the office—and on the other side of the world. When in doubt, humans assume the *status quo,* and, in general, the assumption holds up. The Sun rises and sets once in every twenty-four hour period. Reykjavik and Christchurch remain on opposite sides of the planet. Water continues to boil at 100°C.

Most of these consistencies we take for granted unless we see proof to the contrary. Until the early twentieth century people thought that the continents were fixed in place. Then Alfred Wegener (1880–1930) noticed how well the coastlines of Africa and South America line up. This led to the theory of plate tectonics, which states that huge plates of planetary crust slide across the surface of the planet, moving continents and oceans. Something always assumed to be constant (continents) proved to be changing. Not long ago, scientists believed that matter and energy were fixed; no new matter could be created and no old matter could cease to be. This was believed so obvious that Immanuel Kant (1724–1804) used it as an example of something that could be reasoned *a priori*—that is, before looking at any evidence.[1] We now know that matter and energy are closely related and that one can be transformed into the other.[2] Other examples of newfound change include the movement of galaxies, the origin and extinction of species, and changes in Earth's magnetic field. Sometimes it seems that everything changes.

On the other hand, many things once thought of as unique can now be explained by application of a few simple rules or physical properties. Only ninety-two elements account for all naturally occurring matter. These elements in turn come from combinations of just three types of subatomic particles: protons, neutrons, and electrons. Thus, all known matter can be built with only three kinds of particles.[3] Energy, as well, became a great deal simpler once we realized that four basic forces account for everything: gravity, electromagnetism, the weak nuclear force, and the strong nuclear force.[4]

Scientists, after carefully prodding and occasionally blowing up things, have discovered a few important principles that apply to the universe in which we live. This chapter introduces some of the big ones that have to do with astrobiology. What do we really know about how the universe behaves when we are not looking?

Cosmos

"Cosmos" is one of those words that we use daily without thinking about what it means. The word "universe" refers to the set of all things, but cosmos means something more. The Oxford English Dictionary defines cosmos as "the universe seen as a well-ordered whole." It is a comfortable word and one that assumes we can perceive some order and pattern to the grandness of it all. Before I die, I would like to be able to say I see how life fits into the well-ordered whole. Aiming high, perhaps, but I expect the journey will be entertaining and informative.

Science begins with the assumption that the universe behaves itself. In short, science holds to the idea of a cosmos, both as a starting point—we look for order—and a consequence—we see order. We find that the universe holds fast to some common properties. The laws of physics include a number of propositions and predictions that never fail. The law of gravity always operates, for instance.[5] So does Newton's second law of motion concerning force.[6] Scientists tend to be happy when ideas can be expressed succinctly in numbers.

Just because the numbers work out in one place, does that mean they work everywhere in the same way? What if an experiment done in Reykjavik had a different result than one done in Christchurch? What if the mass of an apple changed when it moved from one city to another? This would be problematic. Not only would scientists be forced to do every experiment locally, the whole idea of common observables would go out the window. I could no longer trust another researcher to give me dependable data. Neither could I go to her with data of my own and ask her to trust them. Luckily, most everything on Earth behaves fairly consistently. It has been discovered that at higher altitudes, gravity is a little weaker.[7] Likewise water boils at a lower temperature in the mountains.[8] These effects are small, however, and careful attention to the environment can make experiments consistent around the world. Scientists rely on this consistency.

Astrobiologists have a problem; many of the events that really interest us do not occur on or near Earth. They occur in the vastness of space, on the surfaces of other planets, and in the hearts of stars. How do we know that the rules we live by apply there as well? To answer that question, we need to travel and do experiments. We start with the assumption that the rules will be the same, but carefully look at results in new places. Mars, for example, has required a great deal of caution. Assumptions about soil chemistry and the makeup of the atmosphere have led explorers astray. Only through careful thought and reconsideration of the data did the researchers come to their present understanding. This kind of experiment and consideration will help us to avoid Kant's mistake—coming to too large a conclusion based on too little experience.

Questions of symmetry and uniformity deal with whether something occurring here and there in space (or time) will occur the same way in both spots. Astrobiology requires a certain amount of symmetry and uniformity to work. We must assume that life on another planet would behave much as life on Earth does. In future centuries, it may be possible to travel to planets beyond our Solar System and try our hypotheses, but in the meantime we have to settle for local experiments—where local means somewhere within a few light hours of our Sun—and trust that the results apply everywhere.

Curiously, the belief in consistency has led people to two opposing positions. Some argue that consistency requires a regulator. God must have created the universe in such a way that it could be understood. Any other kind of God would be cruel. Others claim that consistency means we do not need a God for the explanation. It all happens automatically. The introduction of God to the situation only complicates matters in no need of an extra factor. Both arguments can be refuted by noting that we simply do not have access to information at that level. We do not know why the universe behaves itself. Rather, we reflect that it does. It is unscientific to speculate when no testable hypotheses can be made. Perhaps someday, these questions will be tractable to science. In the meantime, it is simpler to assume that the laws apply everywhere. Why trouble with more assumptions when the ones you have explain everything you see? A surprising number of observations can be explained with remarkably few rules.

We are not mistaken, then, to think that the universe is both well ordered and consistent. A cosmos exists in the place of meaningless space and time. At the very least it can be understood, and it may even hint at more profound truths about the way things operate, if we squint in the right way. Let us look at the details and see where our ideas of consistency and constancy hold up and where they do not.

Transformations and Symmetry

As we deal with cosmological principles, it will be important to define some terms. When a phenomenon (either an object or an event) is changed in space or time, the difference can be referred to as a transformation. The transformation can occur in a number of ways; it may involve moving an object, or changing the direction it points, or even completely inverting it in space. Anything you can do to an object counts as a transformation, but generally we mean changing how it exists in relation to the universe around it. This section introduces transformations so that we can cover how events and objects in the universe behave when transformed. We want to know if they will happen consistently under different conditions, so we need a language for those conditions.

All types of transformation could potentially occur in space and time, though in many cases it becomes difficult to imagine exactly how a time transformation would manifest. Space and time get lumped together in the discussion because, at some level, they are the same thing. This relationship is known as relativity. The speed of light can be thought of as a conversion factor between the two. At this point many people's minds go a little numb. If that applies to you, I will say this: hold on to the idea that space and time relate to one another and that transformations occur in both or in either. There are a number of good books on relativity that will satisfy those who want to know more.[9] Relativity has yet to impinge upon astrobiology and I will not go into it further here.

Four types of transformation concern us: translation, rotation, reflection, and inversion. Figure 4.1 shows a cube that I have folded from a flat piece of paper. Each side of the cube has a number on it. As the paper moves from a flat cross into a three-dimensional cube, note how the num-

LIFE IN SPACE

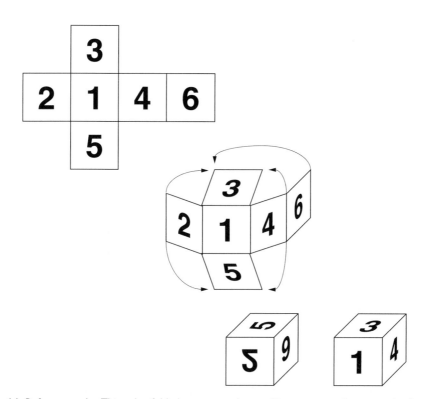

4.1 Reference cube. This cube, folded as you see here, will serve as a reference point for the transformations that follow.

bers end up in relation to one another. Those relationships will help you keep track of the four types of transformation. At the bottom of the figure, I show two different views of the reference cube so that you can recognize it from any angle. If the illustrations seem confusing, you can make your own cube and follow along.

Translation involves moving a phenomenon in space while keeping it oriented the same way. Imagine moving a knight on a chessboard. The knight moves from square to square, but stays facing forward. Translation can occur over any distance in space or any period of time. (The simplest time translation takes place as you read this book. The book's position in

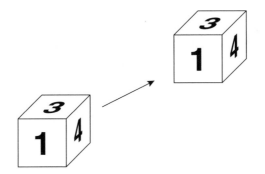

4.2 Translation. The cube moved through space.

space may stay the same, but it moves forward through time.) As we translate the cube in Figure 4.2, the numbers stay in the same places relative to each other and relative to the orientation of the book.

Rotation involves movement of an object or event around an axis. Earth rotates around its own axis and revolves about the Sun.[10] Or you might imagine hiking out to the middle of the desert and taking panoramic pictures of the setting. If you stood in one place and took four pictures—north, east, south, and west—you would have rotated 90° clockwise traveling from picture to picture. Otherwise the events were identical. This is a transformation. Rotation in time is a problematic concept, and will not be addressed here. As we rotate the cube in Figure 4.3, the numbers stay the

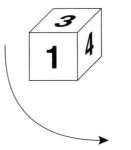

4.3 Rotation. The cube turned on its side.

LIFE IN SPACE

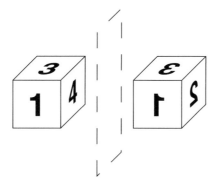

4.4 Reflection. The cube reflected in a mirror.

same relative to one another, but all of them end up tilted on their side relative to us.

Reflection involves flipping an object or event as if to create a mirror image. Your right hand looks like a reflection of your left. Most animals have a line of symmetry such that one side forms a mirror image of the other. For a second example, think of the faucets found on most sinks. Imagine a sink with the hot tap on the left and the cold on the right. The hot tap turns counter-clockwise to start the water flowing. The cold tap turns clockwise. Both taps have the same shape and function, but one is a reflection of the other. The laws of physics prevent us from easily transforming things in this fashion, though it is easy to reflect images and to imagine reflections. When we observe ourselves in mirrors, we unconsciously recognize an important distinction (right and left are switched) and make the spatial corrections in our heads.[11] As we reflect the cube in Figure 4.4, see how the numbers no longer look like normal numbers. Just like text viewed in a mirror, these numbers are now backward. There would be no way to move a cube into this new shape. It has to be changed.

Reflections in space are commonplace; reflections in time, however, can be quite interesting. They require events moving backward in time. In the case of your book, the process might not startle you. The book

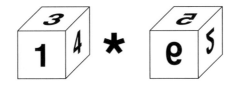

4.5 Inversion. The cube inverted in three dimensions.

moving backward in time looks much the same as the book moving forward in time. In other cases, temporal reflection can be rather disturbing. Imagine dropping a plate so that it shatters on the floor . . . in reverse. The section on entropy, below, explains why this seems so strange to us.

The final type of transformation, inversion, seems even more bizarre. Reflection means flipping an object in one dimension; inversion involves flipping in all dimensions at once. A one-dimensional flip seems straightforward—mirrors do it all the time. We can also imagine flipping in two dimensions. Up becomes down, and left becomes right. (A concave mirror will do this.) In the three dimensions (plus time) of our common experience, however, flipping is far more interesting. Up becomes down, left becomes right, front becomes back, and future becomes past. It hurts the head. Figure 4.5 should make it a little easier to understand. The cube has been inverted in all three spatial dimensions. Take a close look at the numbers. They look a little like the numbers in the reflected cube (Figure 4.4), but not quite. It might also help to compare the inverted cube to the back view of the cube in Figure 4.1. Convince yourself that they are not the same.

A number of objects and events operate in exactly the same way before and after being transformed. In the jargon of physics, they are said to have "reflective symmetry," for example, or to be "space symmetric" with regard to reflection. The terminology may be a bit daunting, but the idea becomes straightforward with a few examples. Some things behave the same regardless of how they are transformed. Some do not. The question is, which are which?

LIFE IN SPACE

Uniformity

Space/time uniformity generally refers to translational and rotational symmetry, and it seems to apply everywhere in the universe for almost all phenomena. In short, no place and no direction in the universe is privileged over any other. Planet-bound as we are, we might be tempted to think that up and down are not reversible, but in the grand scheme of things, that distinction does not matter. The proof of this lies in the roundness of the planet. No such thing as absolute up exists; up always occurs relative to the surface of the planet. Gravity matters because we are so close to an extremely large object, Earth. We assign one direction to be down because of the force exerted by gravity—keeping us next to the planet. So up and down do not seem to have any particular meaning apart from what happens to be nearby.

If life is something that happens to planets, then it should regularly occur within a gravity well. We find up and down tremendously important. Chances are that most life will. Studies have shown that many organisms use gravity to locate themselves. Animals use gravity to aid their motion and plants use gravity to find soil and sunlight. From the standpoint of physics, however, the properties of up and down are local and depend only on mass. They should work equally well on any other planet.

Gravity happens to be one of the best-proven aspects of uniformity. Even before the age of space travel, Isaac Newton (1643–1727) was able to predict the courses of the planets by imagining them as giant masses attracted to one another. As time passes we generate more and more complex models of the Solar System, but those basic predictions stay the same because we have already discovered the most massive objects nearby. Astronomers were able to infer the presence of both Neptune and Pluto based on the effect their masses had on other planets. Gravity turns out to be an important tool for understanding, exploring, and making predictions about objects in space. Space travel only increases our certainty about gravity in other places. We detect more stars, planets, and moons, all of which seem to follow the same basic laws.

The laws work the same everywhere. One of the more staggering discoveries of modern astronomy involves the center of the universe, or lack thereof. During World War II, Edwin Hubble (1889–1953) recognized something interesting: distant galaxies appeared to be redder in color than their closer neighbors. Could it be that distant galaxies were cooler than nearby galaxies? Why might that be? Hubble came up with an elegant solution. He proposed that all the galaxies in the sky were moving away from our galaxy so that their spectra shifted toward red.[12] Subsequent experiments have shown that the model fits well with reality. Galaxies are moving apart—the greater the distance, the faster the movement and the redder the galaxy. Everything moves away from everything else. Compiling thousands of observations, astronomers have calculated an expansion rate for the universe. Galaxies stay the same size, and even associate in clusters, but the distance between the clusters is getting bigger.

We call this observation—of galaxies moving apart at a constant rate—Hubble's law. The speed (v) at which distant galaxies move away from us is directly related to their distance (d) from us.[13] In other words, the Hubble constant, H_o, relates distance to speed according to the equation $v = H_o d$. The universe behaves consistently across time with more and more space coming into being as objects spread further and further apart. Available data place the constant at approximately 71 (km/s)/Mpc.[14]

The idea might seem counterintuitive. Albert Einstein (1879–1955) once said, "What does a fish know about the water in which he swims all his life?" Incredible as it seems, we are not aware of how the universe operates, partly because of its immense scale, but also because it is the very stuff in which we exist. The metaphor I like best for explaining the expanding universe involves a balloon. If we were to think of the universe as two-dimensional—written on a piece of paper, things become simpler. Write the two-dimensional universe on the surface of a three-dimensional balloon, then inflate the balloon. What happens? Everything moves away from everything else. Some cosmologists like to compare the universe to a three-dimensional entity on the surface of a four-dimensional balloon.

If you have trouble with this concept, you're not alone: if the universe is expanding, then what is it expanding into? There are hundreds of questions that this metaphor leaves unanswered. What is the fourth dimension? What is the substance of the balloon? How do we think about the space inside the balloon? How about the space outside the balloon? Are we growing into something? Worst of all—from my noncosmologist viewpoint—if space and time are really the same thing, should we assume time is expanding as well? It causes a certain disquiet. Nonetheless, it explains things quite well—up to a point. If you can imagine a balloon slowly stretching bigger and bigger, you will get the idea of an expanding universe.

One of the real benefits of the expanding universe comes from our ability to rewind. If light takes millions of years to reach us from the most distant stars and galaxies, then we know that as we observe things farther and farther away, we are really looking back in time. If we can look out far enough, we will see some of the earliest events in the universe. The Big Bang, the explosion of our universe into reality, can be inferred by running time backward. If everything moves away from everything else, there must have been a time when everything was closer together. There must have been a starting point, which expanded into everything we know. So we look out farther and farther into space, trying to see things earlier and earlier in time.

Unfortunately, at some immense distance it all becomes a blur. At a distance that corresponds to roughly 13 billion years ago we reach a horizon. The universe appears as a very bright light. We see almost no detail. In every direction we look, if we look far enough we will see the same thing—the flash of energy at the edge of observable time. The balloon had yet to be inflated and was very much smaller than it is now. Some cosmologists argue that it all started as a single point, with no distance in any dimension. This point expanded for 380,000 years until it had spread out to the point we see. Others argue for the universe starting as a cloud, already partially spread out. Either way, every point in the universe was once

very close together. We know that the hot early universe occurred everywhere because we see it in every direction. We know that everywhere
used to be much closer together because it is currently moving apart.

It may sound like a fairy tale. What leads us to believe that the universe
really operates this way? The uniform position and velocity of galaxies and
galaxy clusters around us may be suggestive, but could that not simply
mean we didn't have very good equipment, or that we really are in the center of the universe? It might be tempting to say so. There are additional
data, however, that support the theory of an expanding universe. The definitive evidence comes from what is called cosmic background radiation.

If the laws of physics operated then the same way they operate now, the
universe started out exceedingly hot—so hot that atoms could not form.
Free electrons and atomic nuclei just bounced off of each other and ca-
reened around space. Every time they came close enough to interact, they
were moving so fast that the laws of attraction could not hold them together (in this case the attraction of opposite charges). If everything started
as a single explosion—the Big Bang—then it took 380,000 years for the
universe to cool down sufficiently for the simplest atoms (hydrogen) to
form. So we should expect to see traces of that original furnace in the
night sky, and indeed we do. Background radiation, specifically the presence of microwaves, demonstrates the remains of an initial firestorm that
included everything. This background radiation was first observed in
the mid-twentieth century, and subsequent experiments have shown it to
be remarkably uniform.[15] The Wilkinson Microwave Anisotropy Probe
(WMAP), launched by the National Aeronautics and Space Administra-
tion (NASA) in 2001, has taken detailed measurements of the cosmic
background radiation.[16] The data from WMAP show evidence of very
small irregularities in the earliest visible universe. As matter in the denser
areas condensed, these irregularities probably allowed for the formation of
stars and galaxies.

Uniformity applies with regard to space for as far and as long as we can
collect data. At some point in the distance, we lose track. The universe was

LIFE IN SPACE

so hot that nothing more is visible. Scientists, always looking for the simplest comprehensive explanation, tend to think that the rules must have applied even before that time. If the universe was expanding after it cooled, then it probably expanded before it cooled. After all, expansion could explain the cooling. We know that the farther away we look, the hotter everything gets, and if we look far enough, it becomes a great soup of incredibly hot particles. One way to explain this would be to set up physical laws that changed as you got farther and farther away from Earth. The idea of a Big Bang and an expanding universe is simpler, however—and more compelling—precisely because it does not require us to measure the distance from Earth wherever we are. Physics is not local. The laws apply everywhere.

What we would really like to do is move to another galaxy—millions of light years away—and take the same measurements all over again. If we saw the same thing there, we would be more confident that the same laws really do apply everywhere. Mind you, that trip would take millions of years, so we could not get results back anytime soon, even if we had the technology to take such a journey. While we wait, the simplest explanation lies with a uniform universe.

The Cosmological Principle

All of this brings us to the cosmological principle—no place in the universe is better than any other. Translation and rotation do not affect the way we interact with the universe around us—though in some cases reflection and inversion do make a difference. Everything expands as if from a single point in time and everything exists at that point even though it all gets farther and farther apart. Not only can we displace Earth from the center—and the Sun—but we can say that nowhere at all constitutes the center of the universe. The thought can be rather humbling. After all, it would be nice to know that we had pride of place in the workings of the cosmos. On the other hand, the thought can be extremely comforting. If we do not have pride of place, then there is no reason to believe we cannot understand how the universe operates elsewhere.

As citizens of Earth and residents in the cosmos, we prefer the predictability of uniformity and have every reason to expect that it will apply almost everywhere. It might be worth mentioning here that some localities may not obey the rules. Physicists are quite fond of such places as black holes precisely because they stretch the limits. They redefine what we consider possible. All bets are off for quantum singularities, wormholes, and other irregularities where our working knowledge of space-time simply does not hold up. Call it a travel advisory for the cosmic tourist.

To sum up, the Hubble law and the cosmological principle assure us that the universe behaves itself. We can expect uniformity with regard to spatial translations and rotations. No matter where you look, or in what direction, you will see the same thing, at least at the largest scale. We have also seen, though, that temporal translations may not apply. Your experience with physics does not work evenly for every when, and if you go back far enough, it all gets mucked up in a rather hot soup of particles.

Interestingly, the expanding universe allows us to calculate the age of the universe as well. If you know how fast things move apart, then you know when they started moving apart in the first place. Using the Hubble constant, described above, it is possible to date the origin of the present universe to approximately 13.7 billion years ago. The following 380,000 years were far too hot for any interesting chemistry, or so the cosmologists tell us. If we want to be careful, we should rope off the whole first billion years of the universe as too hot to touch. Subsequent eons should be safer.

Although we want to use caution about time translation, knowing that all times are not equal, we need to remember that most times of interest to us (a few billion years forward and backward) are too small to figure in these types of calculation. Human history can be measured in thousands of years and human evolution in millions of years. Physical laws can be expected to be uniform for the duration of conceivable experiments and observations regarding life. The cosmos appears uniform with regard to translation (no matter where you are or when you are) and with regard to rotation (no matter which way you are looking). This applies across all distances and most of time. Nonetheless, we maintain some humility and know that, on a rare occasion, the truly unexpected turns up.

Having dealt with translation and rotation, we now move on to questions of reflection and inversion. How well does the universe behave with regard to these kinds of translations? Reflection in time raises the issue of entropy, the reason time seems to flow in only one direction. Reflection in space brings up chirality, when an object differs from its mirror image. Inversion may be thought of as four reflections: three reflections in space—one for each dimension—and one reflection in time. Some of the those reflections will cancel each other out, but for the most part, phenomena in space-time can be said to be uniform with regard to inversion only if they are uniform with regard to reflection in all four dimensions, something that appears to be extremely rare.

Entropy

All of us know intuitively that time runs differently forward and backward. Time marches forward; processes that run in one direction simply do not run in the other. This realization has to do with the second law of thermodynamics, which I mentioned briefly in the last chapter. Entropy can only increase or stay the same. In the vernacular of science, entropy measures the amount of order or usable energy in a system. Any closed system will either maintain its organization or lose organization with time. This section goes into a little more detail about how entropy works and what it tells us about time symmetry in the cosmos.

Let us return to our thought experiment in Chapter 3 about the deck of cards. We arrange the cards by suit and number—every card in its place. The deck can be said to have maximal organization because all the cards are in a fixed place. If no outside force acts on the deck, it remains equally organized no matter how long you let it sit. Organization—entropy—remains the same. Now drop the deck of cards from the top of a high building. What are the chances that the deck will stay neatly organized all the way down? (We will assume that they are in no way packaged or constrained.) The chance is pretty slim, though physics says it is at least possible. Gravity, wind, interactions with air molecules, and random interactions between cards will almost guarantee that the cards will separate and rearrange until they land in a chaotic heap on the pavement below. We know that the deck can become disorganized even without the addition of any outside force.[17]

What happens if we shuffle the deck so that the cards are arranged randomly at the beginning? Is there any chance that they will spontaneously rearrange themselves into a neatly sorted deck on the way down? Once again, the possibility exists, but this time it is so rare that it probably would not occur in the lifetime of the universe, even if we decided to devote the rest of human existence to playing with cards. Order does not increase within a system unless work is applied—unless an outside force acts upon the system.

Entropy, the measure of disorder, always increases.[18] You might intervene to organize the deck, of course. In that case the system of cards becomes more ordered. It required work, though, and you expended energy to do it. Whereas entropy—the increase in disorder—applies in the case of the human/deck system, it does not apply solely for the deck. Energy enters the system as human work. For this reason, we must be careful to always define boundaries when looking at entropy. It often seems as though order increases because our field of view is too small.

Another example of entropy can be found in the motion of molecules. Molecules in liquids and gases move about randomly when heated.[19] As they move, they bounce off of nearby atoms and tend to distribute themselves evenly throughout the available space. Think about how smoke will spread out and fill a closed space. The smoke particles randomly interact

LIFE IN SPACE

with the air and, as entropy increases, the whole system moves to the most disordered state. The smoke becomes evenly distributed.

Entropy gets attention in astrobiology because of the extent to which it affects life. If the cosmos tends toward disorder, then we would expect everything to dissipate and decay without the presence of outside work. Organisms use metabolism to fuel work. Metabolism includes all the chemical processes of organisms. In our case, we break down carbon-based molecules to generate energy, manipulate the world around us, and reproduce. The local anti-entropy of life means that some process must be concentrating the energy.

Near human scale, you might think of fire. Carbon molecules in the form of wood and coal are broken down and energy is released in the form of fire. The process cannot run backward; heat and energy cannot be shoved back into carbon chains, reforming smoke and ashes into wood. Time runs in one direction. It is easier to release energy than to store it. Wood and coal come from biological processes that capture solar energy and store it in the form of long carbon chains. When combustion occurs, the carbon chains are broken up into single carbon atoms (which join with oxygen to form carbon dioxide in smoke) and short disordered chains (which make up soot and ash).

At a larger scale, the sun takes form as a giant nuclear reactor that burns hydrogen and helium and larger atoms, destroying mass and releasing lower-energy, lower-order light waves. What seems like the generation of energy really represents an entropic process. Gravity collects matter from vast distances into a single bundle. Some of the mass gets converted into energy and radiated out while the rest remains as a cold, dead star.

On the largest scale, one can say that the entire universe obeys entropy. At the beginning, all energy and matter were packed into a tiny space. Everything was perfectly ordered. As the universe expands and cools, things drift apart, slowly settling into equilibrium. As I mentioned above, the universe appears uniform no matter where you look. Things spread out evenly. At the end of time, physicists expect that the universe will expand so much, and everything will spread so far apart, that gravity will no longer be sufficient to pull a star or planet together. After vast amounts of time, even the rocks and planets will slowly dissolve into a thin gas. Over tens of billions of years, the particles will be spread out over an infinitely large, heatless expanse of space.

This process has been called the "heat death of the universe." For many years, cosmologists debated whether the universe would switch direction at some point, collapsing once again to a very small, very hot point called the "Big Crunch." This would be just like the Big Bang, but backward. In the past ten years, cosmologists have become more and more convinced that this is not an option. Current knowledge suggests that the expansion of the universe will continue until entropy is satisfied.

A simple moral can be derived from all this. Though some physical phenomena (such as objects bouncing off of other objects) can be described equally well with time running either way, time has a direction. Any confused observer would be able to tell which way time was flowing simply by dropping a cup of tea. The cup would encounter a hard surface and become disorganized. By following entropy, he could follow time's arrow in the only direction it flies.[20]

Chirality

We can establish that time is most certainly not uniform with respect to reflection. What about space? Many phenomena do occur the same way in mirror image as they do under normal observation. If we return to the example of a dropped teacup, we would see that the mirror image behaved itself. There are a limited number of phenomena that do not behave themselves, however.

The word "chiral" comes from the Greek word for hand. The right and left hand form mirror images of each other, but are not identical. No matter how you turn them, there is no way to line them up so that they both have the back up and the thumb on the same side. They have the same features in the same order, but they are not the same.

Some molecules are the same way. If the reflected image of the molecule can be superimposed—placed on top of—the original in such a way that they are identical, then the molecule is not chiral. If the process is impossible, they are chiral. The left hand cannot be superimposed on the right, nor the right on the left. This lack of reflective symmetry is called chirality and it comes up often in biochemistry.

Surprisingly few molecules are chiral. Imagine a water molecule, for instance—H_2O. The atoms of hydrogen and oxygen can be thought of as spheres, with the hydrogen molecules a little smaller than the oxygen. They all overlap—two hydrogen spheres stick out of the central oxygen sphere (105° apart) like mouse ears. Flip the image over in your mind. It looks exactly the same. Reflect it in a mirror. It still looks the same. Water is not chiral. Neither are the other major gases in our atmosphere—nitrogen (N_2), carbon dioxide (CO_2), and hydrogen (H_2).

Things become more complicated in solids, but for the most part crystals and rocks are composed of regular arrangements of atoms. At the visible level, they are irregular, but any individual atom forms connections to other atoms in an achiral way. Diamonds, for instance, form when large numbers of carbon atoms link up in a regular pattern. Each carbon makes a covalent bond (sharing electrons) with four other carbons.

Chiral molecules do exist, though, and are particularly prevalent in living processes. Many of the fundamental building blocks discussed in Chapter 16 are chiral (nucleotides, amino acids, some sugars). In each case, the molecule has at least one carbon bound to four different atoms or strings of atoms. A carbon atom attached to four chains, two of which are identical (for example CH_2NOH) would be achiral because you could switch the two identical chains (here the two hydrogens) without changing the molecule overall. On the other hand, when there are four completely different chains, the molecule must be chiral. It cannot be reflected without change. Take another look at the cube in Figure 4.1. It has six different faces arranged in such a way that you know when it has been reflected or inverted. Chiral molecules resemble that cube; only instead of different faces they have different attached chains of atoms.

For many years, physicists assumed that all physical processes were achiral—uniform with respect to reflection. Processes such as gravity and momentum operate without regard to the handedness of the objects they affect. Although none of them relate directly to astrobiology, a number of chiral processes have come to our attention: angular momentum, particle angular momentum, and electromagnetic wave propagation.

For our purposes, we need only say that life lacks uniformity with regard to reflection. Biochemistry often chooses one type of molecule over its mirror image—right-handed sugars and left-handed amino acids. This selectivity comes from the very specific ways that molecules interact and the importance of catalysts in speeding up reactions. Physics, chemistry, even biology tend to be uniform with regard to reflection. This will almost always be the case for macroscopic processes, though some microscopic processes and much of biological chemistry are chiral.

We can now return to our initial question with a little more knowledge. Is the universe well behaved? Yes and no. Phenomena in space are almost always uniform with regard to translation and rotation. No matter which way you look or how far you move, the rules remain the same. Phenomena in space are usually uniform with regard to reflection, and processes that are not can be called "chiral." Chiral processes include many biochemical reactions.

Phenomena in time generally can be thought of as uniform. Time, at least the last 10 billion years or so, appears to be consistent. Notably, the very beginning must have been different in some ways, but as far as we can tell, the next 10 billion years should be more of the same. Rotation in time does not seem to make much sense, since time only involves one dimension—forward and back. Some phenomena in time may be thought of as uniform with regard to reflection; however, entropy ensures that systems and processes always favor one direction—forward toward disorganization.

Inversion, though challenging to think about, can be simplified. Inversion is roughly equivalent to a number of reflections, one for each dimension being considered. So, for a one-dimensional object, inversion equals reflection. For a three-dimensional object in space-time—you and me—inversion represents three spatial reflections and one time reflection. If you need to know what processes behave uniformly with regard to inversion, you must check out all necessary reflections. Sometimes one reflection will cancel out the effects of another. With regard to life, we can see that reflection in time is not possible, so inversion will not come up again in this book.

5

Well-Behaved Observers?

Chapter 4 dealt with the idea that the universe we observe behaves itself. A comprehensible cosmos turns out not to be unreasonable. This chapter deals with a parallel question: "How well behaved are we when we observe the universe?" When humans construct models about how the universe operates, we come with certain assumptions and biases. They may be built in and they may simply reflect the position we hold in the universe, but it has become clear that humans cannot be entirely impartial.

I intend to discuss two ways in which we must be partial: the anthropic principle and the Goldilocks principle. The anthropic principle states that any universe humans observe must contain humans. We could not observe if we did not exist. The Goldilocks principle states that we tend to view ourselves as normal. Whatever diversity exists within the universe, we always tend to view that diversity as centered on our own preferred place. No matter how many beds we sleep in, the one at home will always be "just right."

Even though I have already apologized for a certain amount of imprecision in the introduction, I want to say an extra word here. Both principles, but particularly the anthropic principle, have a very peculiar relationship with cosmology and evolutionary biology. They relate to some of our most fundamental assumptions about who we are, whether we hold a preferred place, and how we observe the universe. Consequently, many scientists hold strong opinions. A large number of competing definitions exist for each concept, and no clear consensus has been reached. Nor, perhaps, will consensus ever be reached. A certain amount of philosophy goes into our judgments about such things, so they may be the proper domain of philosophers. At the same time, these judgments can be foundational to the exercise of scientific observation. So they belong to scientists as well.

This chapter should be viewed as an introduction to basic concepts related to observation bias. In particular, I want to highlight some of the ways our human and terrean biases affect the search for life in space. Many if not most astrobiologists will have their own particular definitions of the anthropic and Goldilocks principles, some of them far more historically informed and logically rigorous. I hope that interested readers will pursue the subject further and explore a variety of possible answers. Here I have provided only enough information to begin that process.

The Anthropic Principle

"Anthropic" comes from the Greek word *anthropos,* meaning human. It contains two major pieces held in constant tension. For the first piece of the anthropic principle, we note that observation requires an observer. This form of the anthropic principle may be poorly named. Any observer would do, regardless of whether or not that observer is human. I will refer to this as a type of sample bias.[1] It is not possible to observe all phenomena fully, nor is it possible to observe from some position outside the universe being observed; therefore, we always deal with limited data. The limited nature of our data must bias our models. So one could say that what we see is a sample of the whole universe. To what extent does any sample represent everything that could be observed? We do not know. Every experiment or observation involves sample bias. (If we could know everything at a glance, we would no longer be making observations.) Astrobiologists want to know how that bias affects statements about the origin, existence, and distribution of life. One of the essential limits to our knowledge about life comes from the fact that we have only one sample available—terrean life.

Terrean life, especially human life, necessitates the second piece of the anthropic principle. One might say that the existence of humans tells us something useful about the nature of the universe. Some philosophers like to ask whether life and humanity are necessary consequences of any possible universe or simply contingent products of this universe. In either case, we know something important. If we are necessary then the universe must be set up with us in mind. If we are contingent, then we represent one outcome among many—which begs the question, "Are we alone?" The second piece I refer to as the specific anthropic principle, because it relates specifically to humans.

You can see how the specific anthropic principle might have important things to say about the validity of observation (critical for scientists) and about the presence of some directing force in the universe (essential to religious thinkers). People have strong opinions. Unfortunately, we cannot completely separate out sample bias from the specific anthropic principle. Humans are the only observers with whom we can reliably communicate. Perhaps in the future, dolphins and parrots will be able to share their cosmologies with us in a way that helps us understand the specifics of our humanity. In the meantime, we can only hold onto the idea that we may be responding to uniquely human circumstances.

The History of the Anthropic Principle

The term was coined in 1974 by Brandon Carter, speaking at a meeting of the International Astronomical Union (IAU).[2] In 1983, Carter emphasized the limitations of his original claim. He urged cosmologists to account for biological constraints in how humans gather data. He also urged evolutionary biologists to account for astronomical constraints on the environments in which evolution occurred.[3] Astrobiology then as now rests on interdisciplinary communication. The anthropic principle was intended to caution astronomers and biologists that they must look at questions of existence, life, and observation as fundamentally related. This very weak form of the anthropic principle, however, was not the first time that sample bias was noted in the case of human observation. One of the best examples comes from the English philosopher of science Francis Bacon (1561–1626). He listed a number of human predilections that color human observation.[4] Self-reflection forces all observers to recognize that their perspective affects their observation.

Whatever the intent of the original term, the anthropic principle picked up a number of differing interpretations in the 1980s. John Barrow and Frank Tipler wrote a book that divided the anthropic principle into several categories.[5] They outlined three types of anthropic principle. The "weak anthropic principle" states nothing more than the impossibility of observing a universe in which carbon-based life could not have arisen. Physical constants, therefore, must have values allowing for the possibility of known life. The "strong anthropic principle" states that the structure of the universe and fundamental constants necessitate the existence of life. In other words, life must occur at some point in the universe. The "final anthropic principle" states that life capable of information processing must occur in the universe and that such life will not cease. Although these distinctions offer some important insights, I think that the specific definitions of Barrow and Tipler conflate carbon-based life, intelligence, and humanity in problematic ways. More recently, Nick Bostrom has written an extensive philosophical treatment of the topic that deals more explicitly with sample bias.[6] This strikes me as far more rigorous in reasoning, though it cannot be applied as easily to astrobiology.

The Fine-Tuned Universe

Some writers and philosophers have made an appeal to what they call the "fine-tuned universe." They note the large number of physical constants that happen to have just the right value to produce a stable universe with intelligent life. The universe fits us just right, and it is possible to argue for a very strong anthropic principle on this basis. Something must have made everything fall into place in exactly the way it did, right?

Physics provides some of the clearest examples. The universe operates on the basis of four fundamental forces—the strong nuclear force, the weak nuclear force, electromagnetism, and gravity.[7] Physicists predict that a universe with only 2 percent difference in strength for at least two of the forces would not have any solid matter. Everything would collapse into one giant lump (too much attraction) or scatter outward into a fine mist of hydrogen with no interactions (too little attraction). In either case, no interesting chemistry could occur. Not only would there be no life, there would not be anything worth studying. Likewise if entropy were only a little stronger or weaker, the universe should cease to be interesting.

Biology also shows us several examples of such fine-tuning. To the best of our knowledge, intricate processes such as information storage and photosynthesis could not operate without the current complex balance of energy usage and structural interactions between proteins. It seems to be a chicken and egg problem. In the case of biology, however, evolution gives us a hint about how the tuning occurred. Gradual changes allow organisms to settle into an optimal arrangement over millions of years. Although the specific sequence of events remains murky, we do know that change happens. The chicken and egg problem can be solved rather simply. Something that was not quite a chicken must have laid an egg from which a chicken hatched. One would still have to ask whether or not it was a chicken egg, but progress has been made. We evade the problem by realizing that chicken-ness (the state of being chickenlike) is not a binary character. In fact, we can imagine chickens, nonchickens, and individuals that live on the edge. These edge dwellers are part chicken but not all chicken. We get into trouble when we create binary categories. The chicken and egg problem, like the biological fine-tuning problem, goes away when we realize we've been trapped by our own definitions.

LIFE IN SPACE

Unfortunately, the physical fine-tuning problem creates more difficulty. We have no reason to believe that our universe evolved from another universe that was slightly different. Where did our universe come from? Are there others like it? Because of indeterminacies in the cosmos defined by quantum mechanics, several influential physicists have supported a theory called the "many-worlds interpretation."[8] In this theory, there exist an infinite number of parallel universes, of which we only experience one. This interpretation suggests that all universes occur, but we only observe the one that has just the right parameters. This would be an appeal to the anthropic principle. We observe the fine-tuned universe because we exist within it, but all the other universes exist at the same time.

Sounds like science fiction, doesn't it? The many-worlds interpretation makes a number of things easier, but in the end it simply appeals to a number of universes that cannot be directly observed. That makes it a question of philosophy that cannot be addressed meaningfully by science —at least not at present. This does not mean it is not true. Rather, we have no way of proving it or disproving it at the moment.

A similar appeal can be made to an intelligent designer. One could say that the universe is finely tuned because someone adjusted the knob. Once again, however, this makes an appeal to something that cannot be directly observed, so we pass beyond the realm of science proper.

The universe appears to have rules almost ridiculously regular, almost ridiculously observable, and plainly favorable to life as we know it. Many people have suggested that we would not be able to make sense of the universe were it not for divine intervention, which connects us in some fundamental way to the order of the universe.[9] More recently, intelligent design proponents have suggested that the universe could not be ordered at all without someone to give orders. Alternatively, quantum physics has opened up the door to infinite universes in which anything can and does happen somewhere. Only our limited perspective stops us from understanding the whole. Several theories exist which extend beyond the current reach of science for an explanation.[10] These theories do not satisfy me, either as a scientist or as a person of faith. They address big philosophical questions, but, before we delve too far, we must decide if they are the right questions.

The Fine-Tuning Observer

We must consider the possibility that the fine-tuning happens at our end, and not in the universe at large. Consider the color-adjustment setting on a television set. The signal is the same, but you adjust the reception. Perhaps the universe obeys no fine-tuned laws. Say instead that the model of the universe created by humans obeys fine-tuned laws. Our limited perspective demands that we make simplifying assumptions. Why should we expect anything other than a well-behaved cosmos? After making an understandable model of the universe, we should not be surprised to find it understandable.

For my part, this version of the anthropic principle makes the most sense. We recognize that our categories form a simplified model of the universe. That model makes sense. That model allows humans to make predictions and affect the universe in desirable ways. That model, though, has been greatly influenced by our unique perspective. It has order because we constructed it that way.

What is our perspective? When we talk about the specific anthropic principle, we can ask quite a few questions about human nature. Humans tend to favor sight over other senses. This suggests that our classifications will lean toward visual characters. For centuries, biologists divided living things into plants and animals based on movement and color—animals move, plants are green. In recent years, biochemical studies have shown us that these can be misleading assumptions. Some stationary organisms are much more closely related to animals than they are to plants. Some plants are white. So, it would seem to make sense to switch our classification system from visual traits to biochemical traits. In particular, biologists favor

DNA (deoxyribonucleic acid, the molecules encoding genes) in the definition of species and larger groups. And yet this may simply replace one bias for another. The deep drive we have to examine things and create an orderly pattern for the universe means we accept simple models to explain things. Humans are quite pragmatic when it comes to making predictions. We like to know how the universe will respond when we do things.

Astrobiologists need to ask in what ways we simplify things. How do those simplifications affect our understanding of life on Earth and elsewhere? We may never be able to eliminate biases altogether, but we can do our best to be aware of them. We already know that we tend to classify things—a bias in itself.

Could there be a downside to classification? Certainly. As we discussed in Chapter 3, the need to classify everything into one of two categories—living or nonliving—has made it nearly impossible to imagine an intermediate stage. And earlier in this chapter we saw that the chicken and egg question rests on a false dichotomy—the idea that it must be one or the other. Binary classification makes it difficult to look for anything that might be somewhere in between—the missing link. We shy away from the uncomfortable examples that mess up our neat classifications, but astrobiology needs to explore some of the in-between cases. When we look at the origins and extent of life, we look at the gray connecting life to nonlife.

The Middle Way

We can reconcile the tension between a fine-tuned universe and a fine-tuning observer by making an appeal to probabilism. Most of us tend to view the universe as "really" one way or another. In that sense we are realists. Another option exists. We could be probabilists. We could say that the universe may or may not be a certain way, but we operate on the assumption that it behaves one way some proportion of the time. It is a matter of consciously separating our model of reality from reality itself.

Recalling Chapter 2, we can connect our confidence or certainty about a proposition with the probability that it will occur. We recognize observation or sample bias. Even if we cannot, in most cases, eliminate our bias, we leave a space for it.

Some philosophers, reflecting on modern physics, have doubts about whether or not a fixed reality exists at all. They ask whether probability distributions—such as those used to understand the propagation of light and the position of electrons orbiting a nucleus—might not be more real than binary operators—the object is there or it is not. They would rather assume a universe, or even multiple universes, filled with possibility.

Other philosophers simply reject our ability to know some things and maintain that a probabilistic model best explains a more determinate universe. I tend to favor this second camp, though I suspect we have now wandered far from the basic purpose of this book: astrobiology. When it comes to looking for the beginnings and edges of life, we must be aware of the limits of our own knowledge. It is important to be able to quantify our confidence in assumptions, observations, theories, and laws about the way life works.

People who are interested in this type of reasoning should check out the literature on Bayesian inference. Scientists and philosophers of science are beginning to get excited about the Bayesian approach. In this version of the scientific method, researchers posit a prior probability (how likely they think a certain hypothesis is to be true) and a likelihood (how well the hypothesis matches observed data). The exciting part of the method lies in the fact that this process can be repeated over and over again. At each stage the prior probability can be regenerated using information from previous stages so that the explanation better matches the data. Just as evolution can produce mechanisms that are more and more efficient, Bayesian inference can produce hypotheses that make better and better predictions. With computers to do the calculation, billions of stages can be conducted in the space of a year.[11]

The Goldilocks Principle

If the human perspective truly matters, then our tendency to favor the familiar may be our greatest bias. Humans tend to classify the world in a way that favors our own position. Politically, we like to view ourselves at the middle of the spectrum. Chronologically, we tend to view others as too old or too young unless they are our age. Scientifically, we look at the world as being more or less suitable to us. Just as Goldilocks declared the first bed too hard, the second bed too soft, and the third bed just right, humans categorize the world in a way that makes our own position seem normal and right.

We utilize the Goldilocks principle in hundreds of ways. The most popular case in astrobiology involves Earth. It would be easy to say that Venus is too hot for life and Mars too cold. Earth, on the other hand is just right. Earth *is* the perfect place to find life exactly like us. But that is precisely the problem. Astrobiologists want to find more than just life exactly like terrean life. We want to figure out all the possibilities and look for life elsewhere. Biologists tend to define life by the biochemical and evolutionary processes we find most familiar. Geologists extrapolate to all other planets on the basis of what we know on Earth. The Goldilocks principle pervades everything.

The popularity of the Goldilocks principle comes from two sources. First, we always build our knowledge outward from the things we already know—the things closest to us. As much as we try, this bias cannot be avoided. It must be a function of how we live and learn. Nonetheless, an awareness of the process can help us to be responsible in how we go about learning. Second, the Goldilocks principle comes from a purely emotional love of familiarity. What we know is comfortable, and, therefore, seems most reasonable. This bias can be much more dangerous.

Astrobiologists have a particular fondness for "extremophiles," organisms that defy expectations by being unlike common life. Biologists look at life on Earth with the presumption that life like humans—thriving at our temperature, pressure, salinity, and so on—must be normal. Everything else becomes an extreme. Just like extreme sports, extreme organisms are unusual and dangerous, right? Surprisingly, no. Many astrobiologists believe that the early Earth was quite hot and the earliest life on Earth thrived in high temperatures (above 60°C). Why should we consider the temperature on Earth now to be normal, but the temperature billions of years ago to be extremely hot?

When we look at other planets, should we look for planets like our own? Many surveys focus on the habitable zone. This zone includes those planets likely to have stable bodies of liquid water, being the proper distance from their parent star. Ice and steam are considered extreme, but liquid water is just right. Certainly this way of looking for life will give us the best chances for success. We look better when we know what we are looking for. In Chapter 8 I discuss what makes a planet habitable for life *as we know it.* Meanwhile, astrobiologists and other scientists continue to expand the limits of what we know, so that one day the search for habitable planets will be less restricted by our biases.

On the popular television show *Star Trek,* Doctor McCoy was known to say about new species, "It's life, Jim, but not as we know it." Astrobiology ponders the question of life not as we know it. Although it may be impossible to separate ourselves from our observation bias, modern research has become quite savvy about prying into the effects. Astrobiologists need to be aware of our anthropic biases, both in everyday scientific practice and at a more abstract philosophic level. Our observations depend on our humanity, and we could never exist in a universe without humans.

Astrobiologists also need to be aware of our tendency to normalize our own position. The Goldilocks principle represents both danger and self-knowledge. We see the risk involved in assuming we are average. We rec-

LIFE IN SPACE

ognize that we are most likely to recognize alien life if it looks like terrean life.

Sometimes the most fascinating questions are the ones hardest to answer. In this case philosophy interweaves with biology, chemistry, physics, and a number of other disciplines. Science is no less useful when it touches on such questions, but it does require caution. Astrobiologists negotiate philosophical and theological issues so as to pick out the questions that can be practically engaged using scientific investigation. Whether we talk about life as it is in an absolute sense or simply our model of life, we can at least say a number of things about what we're looking for.

6

Life in the Cosmos

Life exists in an ordered cosmos. No matter which way we look at the universe and no matter where we go in it, the same fundamental rules apply. Life as we know it also follows certain rules that allow us to see it as part of the ordered whole. In this chapter I leave behind ruminations about possible life and discuss the fundamental rules of terran life.

I'll divide this chapter into the medium, matter, and method of life as we know it. The medium with which we are familiar always involves water, so I will address the special characteristics of H_2O that make it such a popular solvent and transport system. Reactions between carbon molecules, those favorites of biochemists, are the subject of my discussion of matter. I will examine why carbon can participate in so many different chemical reactions and why it makes a good building block for organisms. Finally, I'll discuss the method: how organisms process energy by making, storing, and spending electrons and by creating reservoirs of protons. I catalog the kinds of molecules and chemical reactions that make this possible. By the end of this chapter, you should be able to identify some of the basic properties of life in the universe.

Water, the Medium of Life

Planetary scientists get excited whenever they find liquid water beyond Earth. Water on Mars would be worth investigating, as would water on Europa (a moon of Jupiter) or Enceladus (a moon of Saturn), because water on Earth is so important for life. In fact, astronomers look for planets in the galaxy that are the right size and distance from their star to have liquid water. Meanwhile, biologists are fascinated by saltwater, which stays liquid below 0°C, and water under high pressure, which stays liquid above 100°C. Organisms with adjusted biochemistry on Earth can live in these extreme environments. In fact, everywhere that we find liquid water on Earth, we find life.

Conversely, life has little tolerance for dehydration.[1] Water forms a necessary component of all living cells. It accounts for around 50–60 percent of mass in humans and is the largest component in most organisms. Land and sea organisms have complex mechanisms for keeping it in. Freshwater organisms have complex mechanisms for keeping it out.

What makes water so important? It forms the basic solvent for the chemistry of life. Liquid water provides a stable environment for chemicals to move and interact with one another. The water molecule (H_2O) has one oxygen and two hydrogen atoms. Although the three atoms share electrons, the oxygen atom has a stronger grip than either of the hydrogens.[2] This inequality makes the water slightly polar; the oxygen has a slight positive charge and the hydrogens have a slight negative charge. The polar atoms attract ions and make water a perfect solvent for ionic compounds such as salts. Common table salt (NaCl) is a good example of this. It also means that ions can be suspended in a usable form surrounded by —but not too strongly attached to—water molecules. Figure 6.1 shows the weak polarity of water.

Life depends on the maintenance of fairly specific concentrations of ions such as sodium and potassium. Organisms need to be able to access

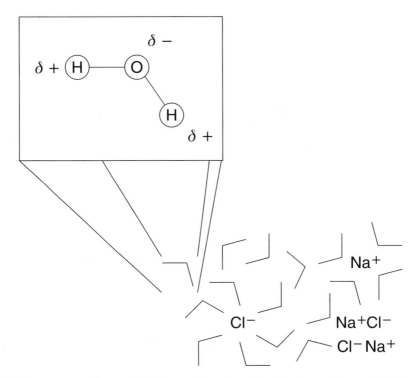

6.1 The structure of water. This diagram shows how the slight charge ($\delta-$ or $\delta+$) allows water to dissolve salts like NaCl (table salt).

the ions for reactions. In fact, the regulation of ion concentrations forms one of the most important mechanisms for controlling biochemical reactions. Thus the ability to dissolve and hold ions is necessary for any medium in which life exists.

At the same time, life needs a good solvent for organic compounds. Because so much of biochemistry involves carbon-based molecules, the solvent for life will need to support them, too. Water may be polar, but it is only slightly polar, allowing it to dissolve some small organic molecules. Long carbon molecules, such as the lipid chains discussed in Chapter 16, orient themselves to avoid water. As the saying goes, oil and water do not mix. Smaller organic molecules, such as simple sugars and nucleic acids, can be dissolved in water, but only when alone. Once formed into long chains, they separate out. The Egyptians built huge pyramids by stacking small stone blocks, one on top of another. Biological structures operate in much the same way. Huge proteins, starches, nucleic acids, and fats (some of which repel water) can be assembled from small pieces transported through the aqueous environment.

Water provides an important environment for life, but it is not the only such solvent. Several small organic compounds have similar properties. We might imagine life existing in a methanol-based environment, for example. Methanol, ethanol, sulfuric acid, and acetic acid all have some of the polar properties of water and have been suggested as possible solvents for life. Ammonia also presents interesting possibilities.[3]

Polarity, however, is not the only characteristic of water that makes it a good solvent for life. We should also consider the fact that every water molecule can dissociate into a positively charged hydrogen (H^+, a cation) and a negatively charged hydroxide (OH^-, an anion). The dissociation of water makes it an ideal reactant for many biological reactions. One of these, a dehydration reaction, involves two molecules joined together by the exclusion of water. A hydrogen from one molecule joins with a hydroxide from another molecule to form a water molecule while the two original molecules link together. A dehydration reaction is shown in Figure 6.2.

This simple reaction occurs in the formation of most major organic macromolecules. The nature of these molecules will be covered in detail later, but for the moment it is enough to say that dehydration reactions are incredibly common in biochemistry. Organisms string together simple units (monomers) into long chains (polymers) with important biological functions. Polypeptides (forming proteins), nucleic acids (for data storage), and carbohydrates (for nutrition) all require dehydration reactions.

Dehydration, of course, can be reversed. The reactions that string monomers together can be reversed to pull them apart again. Life can be so badly damaged by acids and bases precisely because the excess hydrogen

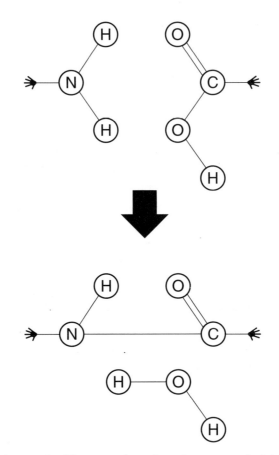

6.2 A dehydration reaction. The atoms shown here demonstrate the joining of an amine group to a carboxylic acid group. Water forms and the two groups join together.

and hydroxide readily react with biological molecules, destroying them by pulling out oxygen or more hydrogen to form water. Cells need to regulate their acidity, or more formally, their concentration of H^+ ions, in order to protect their molecules. This second property of water seriously limits our list of possible life solvents. Only methanol and ethanol share water's acidity and polarity, being neither too acidic nor too basic. Ammonia and hydrogen sulfide have similar acidity, though neither is quite so polar as water. Sulfuric acid, acetic acid, and hydrogen chloride are all too acidic, shedding hydrogen ions too readily.

A third important character of water relates to the range of temperatures at which it remains liquid: above 0°C and below 100°C, although those figures are dependent on pressure. This range, though not the widest among the common solvents, is relatively broad. Carbon dioxide, for example, only forms a liquid between −79°C and −78°C. For life to develop and thrive on a planet, we expect that it would need a constant environment for millions if not billions of years. Maintaining temperature in a range of only 1°C for so long would be difficult. On the other hand, 100°C seems like a reasonable range, allowing a planet to have a slightly irregular orbit with liquid water over large areas. We can easily imagine a habitable zone of liquid water forming a broad territory around many stars. Not only does Earth have abundant and stable liquid water, but liquid water might occur in the martian subsurface and in pockets around some satellites of Jupiter and Saturn.

Although carbon dioxide doesn't seem like a possibility, several other solvents remain liquid over an even broader range of temperatures than water. Methanol spans 163°C (−98°C to 65°C), acetic acid, 111°C (17°C to 118°C), ethanol, 192°C (−114°C to 79°C), and sulfuric acid, 280°C (10°C to 290°C). Hydrogen chloride (29°C), hydrogen sulfide (23°C), and ammonia (44°C) all have small ranges in comparison. Some astrobiologists suggest that the range from 0–100°C contains most of the biologically interesting biochemical reactions. It remains to be seen, however, whether this causes water's suitability or results from it. If organisms only exist in the range of liquid water, then, in accordance with the anthropic principle, we would expect almost all biochemical reactions to occur there.

Finally, and perhaps definitively, water occurs abundantly. Carbon, nitrogen, and oxygen all form within the core of stars like our own Sun. Heavier elements such as sulfur and chlorine require hotter stars. We find water throughout the Solar System because it is composed of only three atoms. It possesses one relatively abundant oxygen atom and two ubiquitous hydrogen atoms. Only methane and ammonia have similar ease of formation.

Habitable zones could be constructed for any solvent's liquid range in just the same way that they are constructed for water, but water appears to be a good fit for terran life for a number of reasons: polarity, acidity, reactivity, liquid range, and abundance. Water is such a perfect fit that astrobiologists frequently determine a habitable zone around a distant star by looking at the proper range for planets that maintain the right temperatures for liquid water over long periods of time. Alternative life chemistries, though speculative, do present other possibilities for a habitable zone. We must remember, however, to match the appropriate chemistry to the appropriate zone.

Carbon, the Material of Life

Biochemistry as we know it is organic chemistry; it involves reactions between carbon molecules. To be more precise, twenty-six different elements make up all known organisms, but around 95 percent of all living matter consists of only six elements: carbon, hydrogen, nitrogen, oxygen, phosphorus, and sulfur. These six elements are so common that many astrobiologists simply refer to them as CHNOPS.

The six basic elements can be thought of as a set of blocks that can be assembled in a huge variety of ways. The molecules that they form act as building blocks for even larger building programs, scaling up through organelles (mini-organs inside cells), cells, membranes, organs, and organisms. A more familiar way to think about it would be to compare the elemental building blocks to letters, which join together to form words, which join together to form sentences, paragraphs, books, and even libraries.

In life a few simple units come together in nearly infinite ways. At every level of life, this combination and recombination of a small number of simple pieces multiplies the possibilities; the number of possible interactions and pathways for life takes on dizzying proportions. At the level of organisms and ecosystems, it may seem impossible to understand life in all

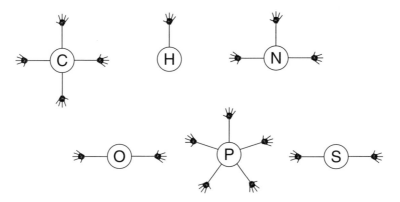

6.3 Basic building blocks—the atoms used to construct life.

its complexity and diversity. Intelligence and behavior, in particular, can be very difficult to grasp. In science, therefore, we use reductionism—the idea that events can be explained by interactions at some lower level of complexity—to tackle these problems. And so we will now look at the most basic units usually addressed in biology, atoms. Some physicists and chemists are starting to explore the implications of quantum theory on life, but at present these efforts are very limited.

If we think of CHNOPS as our basic set of blocks, then atomic chemistry can be surprisingly simple. The number of electrons held in the outer shell of each atom dictates the way it interacts with other atoms. For right now, it is not essential to know exactly how this works. I will say that those outer shell electrons can be shared between atoms to form covalent bonds. The covalently bonded atoms are called molecules. In fact, any set of bound atoms—from a single atom existing alone to carbon chains thousands of atoms long—can be referred to as a molecule. So, for this chapter, let us think of atoms as blocks and molecules as the structures built with them. Each block can be thought of as a sphere with metaphoric hands that can connect to other blocks, as shown in Figure 6.3.

Collections of covalently bound atoms make molecules, and those molecules found in living systems are called "biomolecules." Carbon acts as

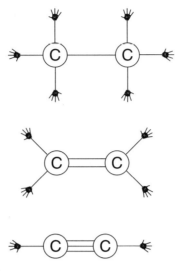

6.4 Carbon-carbon bonds.

a skeleton, forming long central chains and joined loops to which other elements are attached. Nitrogen, oxygen, phosphorus, and sulfur can be found either replacing one of the carbons in a chain or connecting in functional groups on the outside. The functional groups allow the chains to interact with other molecules. We have already discussed what happens when a molecule has an extra hydrogen or hydroxide that can break off and interact. Other common functional groups include amine (NH_2), commonly used in bond formation, and phosphate (PO_4), commonly used for energy storage. These elements can also be used to change the bond angles or to change the way the molecule holds electrons. Hydrogen, meanwhile, fills in the gaps by binding to the molecules wherever an open space is available.

The carbon atom has four hands, making it one of the most versatile in making chains (Figure 6.4). Two carbons can shake hands to form a single bond (C−C); or shake two pairs of hands, a double bond (C=C); or shake three pairs of hands, a triple bond (C≡C). The three types of bonds are not identical. Triple bonds are the strongest but also the shortest; single

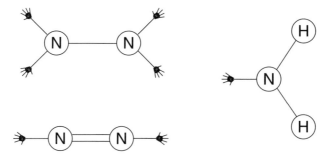

6.5 Nitrogen-nitrogen bonds. An amine group appears to the right.

bonds are the weakest but also the longest. Perhaps most important, because shared electrons form the substance of the bonds, the three kinds of bonds hold electrons in different ways. Below, I discuss how electrons influence the properties of organic molecules and define the method for life. Double and triple bonds tend to allow electrons to travel more freely between atoms. Thus, the ability to form multiple bonds contributes to the usefulness of carbon in making chemically active and interesting molecules.

Nitrogen, one column over from carbon on the periodic table, has one less electron available for bonding. Each of these atoms typically forms three bonds to other atoms. Like carbon, they can form double bonds (Figure 6.5). Two nitrogen atoms will frequently make a triple bond ($N \equiv N$). This makes nitrogen gas, the most common molecule in Earth's atmosphere. Because nitrogen atoms only have three hands, the triple bond means they cannot bind to anything else. Nitrogen gas, therefore, reacts relatively rarely with other molecules. The nitrogen atoms bind together too tightly. Incidentally, this means that most organisms cannot get their nitrogen from the atmosphere. Special biochemical pathways are required. Nitrogen atoms frequently occur in organic molecules as one atom in a circle of carbons, or as a side group on a long carbon chain. Three bonds mean that a nitrogen atom can form one double bond and one single bond, giving nitrogen some of the same electron-sharing properties as carbon. Some nitrogen-containing molecules, called amines, show up regularly in biology. Familiar examples include vitamins (from "vital amines") and amino acids.

Oxygen forms two bonds and commonly occurs as gaseous O_2, the part of air essential to respiration, and H_2O, water. Oxygen can also act as a bridge, connecting two chains or replacing one atom in a ring of carbons. The dehydration and hydration reactions mentioned in the section on water are central to the way terrean biochemistry operates. Oxygen and hydrogen account for almost 90 percent of the atoms in the human body, simply because so much of the human body consists of water. If we leave the water out, however, then the remaining molecules consist almost entirely of carbon backbones with the occasional oxygen tacked on.

Phosphorus—one period lower on the periodic table than carbon, nitrogen, and oxygen—forms bonds using more distant electrons. Those electrons are less tightly bound to the central atom, making the bonds more flexible but less strong. Phosphorus atoms tend not to bind to one another but to shuffle their electrons, making three, four, or five weaker bonds to other atoms. Perhaps the most common use of phosphorus in living organisms comes in the form of a phosphate group, which appears in Figure 6.6. The phosphorus atom shakes hands with four different oxygen atoms in a way that allows two of them to make further bonds outward while distributing several electrons over the whole structure. Phosphate groups are notable because they connect sugars to make up the DNA helix.

Sulfur lies next to phosphorus on the periodic table and has some of the same properties. It can form two to four bonds. The last and least common member of CHNOPS, sulfur rarely occurs as an integral member in a long chain of molecules, though it can act as bridge, in a fashion similar to oxygen. Sulfur is more likely to be found as a side chain in amino acids and signaling molecules.

Hydrogen is the simplest element. It occurs abundantly in the universe and in biochemistry. Having only a single electron, hydrogen can only

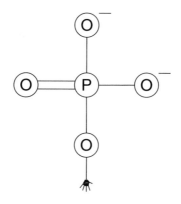

6.6 A phosphate group.

form a single bond, and caps off the spare connections in almost all biomolecules. If you think of atoms as a set of building blocks, then hydrogen would be used for the façade. It smoothes out edges and ties up loose ends. Because of this, hydrogen occurs abundantly in biological molecules. Some estimates place it at one tenth to one fourth as common as carbon (not counting the hydrogen in water).

A large number of other atoms are extremely important to life as well. I have focused on CHNOPS because these are the atoms that bond covalently to make the molecules we define as biological. A number of other atoms bond ionically.[4] Calcium, potassium, sodium, and chlorine are among the most common. Sometimes these essential atoms associate with a carbon backbone. Often a single rarer element will be held by a complex ring of carbons and nitrogens bonded together with many double bonds so that they can share electrons. The ability of the complex molecule to share electrons with the central atom results in unique and important biological functions. Heme, which carries oxygen in the human bloodstream, has such a carbon ring associated with iron. Chlorophyll, the central molecule in converting light into chemical energy (phototrophy), has a similar ring holding magnesium.

Another use for the rarer elements involves concentration gradients.

Organisms frequently send signals and store energy by shifting ions from one chamber to another. That, however, is a subject more appropriate to our next topic, biological energy. Before addressing that, I would like to say a few words about alternative elements.

Alternatives to Carbon

Do we have to use carbon for the backbone, or would something else work? Astrobiologists not only investigate how things are, but also consider possible alternatives. Carbon forms the backbone of all biomolecules on Earth, but does that mean that life could not be based on a different atom elsewhere? "Organic" chemistry has come to mean the chemistry of carbon, but might there be inorganic biology?

We can imagine a number of other atomic constituents for life, just as we could think of a number of other solvents. Still, the one we have on Earth seems to be the best as far as we can tell. One of the primary benefits of carbon comes from its ability to form four separate bonds. This ability means that carbon can form into long chains with side groups and multiple rings and still have extra electrons left over for double and triple bonds. (Remember that multiple bonds can be important for electron sharing, which will be important later.)

Since the ability to make four bonds is so useful, speculative biologists (and science fiction writers) have been attracted to silicon (Si), located directly below carbon on the periodic table. Silicon has an extra set of electron orbitals, but the highest orbital is the same shape as the highest orbital in carbon (called a *p*-orbital), and the outside shell is also half full (four out of eight possible electrons). Silicon can form four bonds and, in addition, seems to be extremely abundant on rocky planets like Earth. Though it is not as abundant as carbon in the universe at large, it is far more common in the rock of planets like ours.

The major objection raised regarding silicon-based life comes from the weakness of Si—Si bonds. Atoms may have a large number of electrons, but it is the electrons in the outermost shell that determine the strength and character of covalent bonds. Because silicon has more electrons than carbon, the bonding electrons are farther away from the nucleus and, thus, the bonds they form are weaker. In a water environment, silicon-based life simply would not hold together very long. This does not mean, however, that silicon life is impossible. It means that silicon life would require a different solvent. William Bains, a systems biologist, has suggested that silicon life would be feasible in a liquid nitrogen environment.[5]

Other alternative biochemistries could be based on nitrogen or phosphorus. The first suffers from limited bonding possibilities (three instead of four) whereas the second suffers from the same problems as silicon (weak bonds, less common multiple bonds). At present, no options suggest an easily detectable variant on life. Although silicon life in liquid nitrogen remains feasible, it is not clear that liquid nitrogen environments could be easily found or explored. Carbon-based life promises to be the only game in town for the foreseeable future.

Electrons and Protons: The Method of Life

Now we know two very fundamental things about life—terrean life at least. Life happens as a form of carbon chemistry and life occurs in water. Yet if we were to stop there, I think we would still be a little short of what most people consider life. After all, a great deal of carbon chemistry goes on in water and no one gets excited. Every time you take an antacid, you are dissolving carbon in water. Even if we limit ourselves to organic chemistry (with carbons covalently bound together), we still have a wide variety of meteorites and ocean chemistry as well as a great deal of oil just sitting around. Oil, of course, represents living organisms that have died and been chemically altered over millions of years. When we think of life we expect something more dynamic. Even the most sedentary of organisms grow, divide, and acquire resources. Despite all the philosophical talk about a formal definition, anything that cannot act in some way on its environment will make a poor candidate for life.

Chemistry provides a number of insights into the way organisms operate. I could say simply that all biochemistry involves shifting electrons about, but that could be said of all chemistry. Most chemical reactions occur when an electron gains or loses energy, associates with or disassociates from an atomic nucleus, or interacts with other electrons. What is so special about biochemistry?

The simple answer is not very satisfactory. Nothing. There does not appear to be a significant difference between biological chemistry and abiological chemistry, between how molecules interact within organisms and how they interact outside organisms. Sometimes it can even be very difficult to decide where an organism leaves off and its environment begins.[6] Chemically nothing seems to be special about life. Reactions that occur in living cells are the same reactions that occur everywhere else.

The simple answer, though correct, is not the whole story. Any amateur biologist can tell that something special happens inside living cells. I see two reasons for this. First, organisms have improbably high concentrations of certain chemicals in very complex and important ratios. It seems unlikely that the right number of fats, sugars, proteins, and nucleic acids would spontaneously come together to form a human. We must presume, therefore, that humans (indeed all organisms) have some mechanism for maintaining their chemical makeup. In other words, living things do not undergo chemical reactions indiscriminately. They have mechanisms for encouraging some reactions and discouraging others, for separating things that need to stay separate, and for avoiding some chemicals altogether.

Second, biological reactions often move much more quickly than abiological reactions. All reactions are ruled by entropy. Energy must be lost. If a great deal of energy is lost, then the reaction will occur rapidly. Vinegar and baking soda, for instance, react immediately, heating up and releasing gas.[7] The molecules at the end have less energy than the molecules at the beginning, so entropy forces the reaction forward. In this simple case, the process occurs easily because the reaction just slides downhill from one state to another. Imagine, though, that the atoms had to get shaken up before they reacted. In the case of fire, the end products have lower energy. They must, because heat is released. The fire, however, cannot occur as spontaneously as the vinegar reaction; it requires a spark of energy to get things started. In order to change chemical structure, a set of atoms may need energy to be put in before energy can come out, just like pushing a cart over a hill and into a valley.

Molecules in nature bounce around. Heat energy ensures that all atoms move, if ever so slightly. Molecules in gases move quickly, molecules in liquids more slowly, and molecules in solids slower still, but all atoms move. Many reactions will occur when random energy from the environment bounces them uphill to an intermediate state. They can then fall back down the same way or go downhill on the other side. In this way, there are a great many abiological reactions in nature that occur improbably, but do occur. Complex carbon chains, such as those in proteins and nucleic acids, form in open space and inside meteorites based on this process, but only rarely. If you have enough carbon atoms bouncing around, some of them are bound to react with one another and form more complex molecules. Life distinguishes itself by its ability to make these reactions occur at predictable times and places so that they concentrate and apply energy.

The question then becomes this: In what way does life collect, store, and apply the energy? How do organisms maintain the right balance of chemicals, speeding up some reactions and slowing down others? Life has two principal methods, redox chemistry—passing high-energy electrons from one molecule to another—and chemiosmotic potential—building up high concentrations of ions.

Passing Electrons, or Redox Chemistry

The primary way that organisms acquire energy involves passing electrons between molecules, otherwise known as redox or oxidation-reduction chemistry. Oxidation occurs to a molecule whenever it loses an electron.[8] Reduction occurs when a molecule gains an electron.[9] When one molecule oxidizes another, the first molecule is reduced. When one molecule reduces another, the first molecule is oxidized. It is all about the passage of electrons.

Entropy increases, so redox reactions have to lose a small amount of energy. Some molecules are said to be more reduced than others; they have more potential energy and more potential to do work. When one molecule reduces another, the second molecule must have less potential energy than the first. Some biological mechanisms involve passing a single electron through ten or more different molecules, with work being done (and energy being lost) at every stage. Still, nothing particularly biological has been mentioned.

Biology happens when we recognize the universal ability of organisms to sequester energy acquired from the environment. One version of this process, quite possibly the oldest version, occurs in chemotrophs. Chemotrophs use naturally reduced fuel atoms like sulfide (S^{2-}) and iron (Fe^{2+}), or more complex fuel molecules like glucose. The organisms steal electrons and oxidize the fuel. The stolen electrons then pass through a series of biomolecules, doing work. At the end of the process, the spent (low-potential) electrons get dumped back into molecules that are expelled as waste.

A second version of energy acquisition occurs in phototrophs, organisms that use sunlight for power. These organisms collect their energy from the Sun by using photons to promote electrons. Low-energy electrons in complex biomolecules like chlorophyll get promoted to high-energy electrons by absorbing photons. Chlorophyll then acts to reduce another molecule and starts a redox chain. Once again, the molecules do work in the process. Phototrophy can occur in a number of ways, even within a single cell, so general rules can be elusive. Often, however, the spent electron gets returned to a chlorophyll so that it can be promoted again. At other times, electrons are taken from and returned to the environment.

In plant phototrophy electrons are stolen from water ($2H_2O$), forming oxygen (O_2) and hydrogen ($2H_2$). This produces the oxygen that plants release into the atmosphere. Plants consume water and generate oxygen. At the same time, high-energy electrons are found or made and then used to reduce biomolecules and do work.

One real benefit of modern biochemistry comes from the ability to store reducing potential. In theory, the earliest organisms could only use their electrons immediately, but over time systems arose that could store high-energy electrons for future use. Rather than squeezing all the energy out of an electron immediately, the organism could reduce a stable molecule that would hold onto the electron and use it for reduction at a later time.

Terran organisms store reducing potential in several common biomolecules called electron carriers. The most fundamental of these is adenosine triphosphate (ATP), and the next two most prominent carriers are nicotinamide adenine dinucleotide (NAD) and nicotinamide adenine dinucleotide phosphate (NADP). We'll put aside the discussion of ATP for later, and turn first to NAD and NADP.

Adenine and nicotinamide are both nucleotides, the basic units from which genes are constructed. (See Chapter 15 for a full description of nucleotides and other fundamental biomolecules.) Bonded together, they make a convenient electron carrier. The nicotinamide has a ring made up of five carbon atoms and one nitrogen. As I mentioned above, nitrogen and carbon have the ability to share electrons in a variety of ways. In this case, the ring has a number of shared electrons and remains stable in two different formations. In the first formation, NAD^+, the oxidized ring carries a positive charge on the nitrogen and has four hydrogen atoms attached.[10] In the second formation, NADH, one electron and one hydrogen have been added so that the nitrogen has a neutral charge and the carbon across from it in the ring has an extra hydrogen (five total). The hydrogen comes on and off easily, taking the electron with it. Two of the most common metabolisms, phototrophy (as in plants) and respiration (as in animals), produce charged NADH. The charged molecule can then be transported to various parts of the cell and used to power reactions.

NADP is NAD with an additional phosphate group added near the adenine. It functions much the same way as NAD, having two formations, $NADP^+$ and NADPH. NADPH has slightly more reducing power than NADH, making it harder to form and more powerful.[11] I would be remiss not to mention that there are a number of other electron carriers as well, though I will not describe them in detail here. Common examples are flavin adenine dinucleotide (FAD/$FADH_2$) and ubiquinone.

But wait, there's more! The principal redox chains in chemotrophy and phototrophy do more than just pass and store high-energy electrons. They also herd protons across membranes. Protons can be thought of as hydrogen atoms missing an electron (H^+); they have a positive charge. The positive charge makes them ions and the pumping means that one side of the membrane has more protons than the other. The buildup of any ion on one side of a membrane produces a different kind of potential energy—in this case, chemiosmotic potential.

Storing Protons, or Chemiosmotic Potential

Chemiosmotic potential is a little complicated, so I will ask you to bear with me for a moment while we put the pieces together. First we need a reservoir of water with a semi-permeable membrane dividing it into two parts. A semi-permeable membrane allows some things through but not others. In this case, the membrane allows water through, but not ions like sodium, chloride, and larger molecules. Sodium and chloride are fairly common, so we will stick with them for the moment.

Imagine that table salt (NaCl) has been added to the water on the right side of the membrane. Water on the left remains pure. What will happen? The salt concentration on the right exceeds the concentration on the left; consequently, the sodium and chloride ions will try to spill over into the left side.[12] Or, if we look at it another way, the water is more concentrated on the left, being mixed with nothing. The water, then, will try to spill over into the right so that the concentration of water will be the same on both sides.

It seems a little silly to think of water or salt "wanting" to do anything. I

LIFE IN SPACE

apologize for the anthropomorphic language. Later on, it will be incredibly important to distinguish between organisms with intentions and objects that simply settle into low energy positions as a result of physical laws. In this case a full explanation of the physical laws would be too time-consuming and strictly nonintentional language sounds very dry. When I say the water wants to get to the other side, I am describing a probabilistic process somewhat colorfully.

We call the force of the water pushing to get through the membrane osmotic pressure. Since the membrane is permeable to water, more water will end up on the right-hand side and the membrane will bulge out to the left. We call the force of ions pushing to the left—the force of water pushing to the right—chemiosmotic pressure. Chemiosmotic potential arises when the ions cannot pass through the membrane. Potential energy gets stored up and that potential energy can be used to do work.

Now imagine that the membrane contains tiny gates that will allow the ions through, but only at a cost. In cell biology, the process involves biomolecules that change their shape when the sodium and chloride pass through. Perhaps the shape change causes another molecule to be trapped or released. Perhaps the change sends a signal or encourages a specific chemical reaction. All of these scenarios play out in terrean biochemistry.

Although I find the use of ion gates fascinating in general—indeed the regulation of sodium going in and out of neurons appears to be central to how we think—the question of energy storage remains before us. How do living cells store energy? They create ion gradients. They concentrate ions in one location by pumping them across a membrane and not allowing them back. They store energy as osmotic potential.

The example most essential to terrean life involves the pumping of protons into membrane chambers. Almost all organisms generate energy using some form of redox chain that pumps protons across a membrane. The protons can only return by means of a gate called ATP synthase. As the name suggests, ATP synthase synthesizes ATP. Like NAD and NADP, ATP starts as a nucleotide. The base group, adenosine, may have a single phosphate group attached (adenosine monophosphate or AMP), or it may occur with two or three phosphates attached one on top of the other. These arrangements make adenosine diphosphate (ADP) and triphosphate (ATP), respectively. The phosphates can break off, releasing energy that can be used in other reactions. ATP has the greatest charge of the three, and one or two phosphates can be broken off at a time. Most commonly, organisms oxidize ATP to form ADP and do work (although ATP/ AMP and ADP/AMP reactions occur also). All known terran organisms live by the production of proton gradients, storing energy as osmotic potential, and the subsequent charging of ATP. NADH and other electron carriers may vary, but ATP appears to be universal. I describe the whole process in detail in Chapters 15 and 16, but for now, just keep in mind the central role protons and osmotic pressure play in life.

Getting Enough Energy

Having said that life does all of its work on the basis of charged electrons, we must explore what this says about the possible sources of charged electrons. The most abundant source of energy on Earth is solar radiation. Phototrophy, the conversion of light energy into chemical energy by means of biochemistry, currently accounts for at least 99 percent of all the biological work on the planet. All other sources are almost negligible if we look at life on a planetary scale. Phototrophy produces higher-energy electrons than any other form of energy acquisition, making it easier to do work. At the same time, the influx of solar radiation is constant and abundant over a period of billions of years. Chemotrophs acquire energy from available reduced compounds. For the most part, this means that they either eat phototrophs or are dependent upon phototroph waste. Energy enters the system by way of sunlight and then works its way through several organisms.

A small subset of the chemotrophs pick up electrons from reduced inorganic compounds. These organisms are known as chemolithotrophs and depend upon molecules like hydrogen (H_2), methane (CH_4), iron (Fe^{2+}), and sulfur (S^{2-}) that have become reduced as a result of geological processes. Volcanoes and deep-sea vents release large quantities of such molecules. Chemolithotrophs live off of this energy. Astrobiologists think that the early Earth, as well as a number of celestial objects in our own Solar System, may have been able to support a small biosphere using only geological energy.

To put it all together, we now have a simple picture of life. Based on life on Earth, we can make several assumptions about the fundamentals of life as a general phenomenon.

Life as we know it can be found only in an aqueous environment. Water's ability to act as a solvent for diverse chemistries over a wide range of temperatures makes it a unique medium. Hydrogen and oxygen occur abundantly in the universe, both together and apart. Within our stellar system, there is at least one region, and possibly more, where liquid water can be maintained for millions of years.

Life as we know it can be constructed from only one basic material. Carbon atoms can create four strong covalent bonds to other atoms. Chains of carbon atoms form the basis for just about every biological molecule, though a small suite of other atoms accentuate the basic structure. These chains can form into rings and complex structures that allow them to hold electrons in common and use them as energy collection devices (as with chlorophyll). Carbon also occurs relatively frequently in the universe.

Life as we know it always functions by use of redox chemistry and proton gradients. Life uses reduced compounds as a way to store and move energy around. Work can be accomplished by the use of charged electron carriers that hold high-energy electrons. These carriers may be created directly or generated using the potential energy from collected protons.

We could certainly imagine other media, materials, and methods for supporting life, but which combinations of the three are feasible and likely to occur in a detectable location? It's an open question for both chemists and planetary scientists. Until such a time as alternative chemistries can be explored, however, the search for life as we know it will dominate the field.

So astrobiologists have a definite program when looking for life in space. The environment must be rich in atoms heavier than hydrogen, therefore we are most likely to find life near a star. Earth-like life requires a liquid water habitat in which chemistry can occur. This limits us to a sphere around the star in which solar heat can maintain liquid oceans but not boil them away (the habitable zone, discussed further in Chapter 8). Or, instead of solar heat, geothermal heating may result in liquid water habitats in other locations. Earth-like life requires a few basic elements, CHNOPS, to be present in sufficient quantities, suggesting that life would occur on a rocky planet or moon. Finally, there must be an influx of energy, either from radiation such as sunlight or from some source of reduced compounds such as reduced iron from geologic activity. In short, we are looking for a rocky planet or moon close, but not too close, to a parent star.

7

Life among the Stars

We now turn to the question of where life in space may be found. It requires a fairly precise balance of energy, so we would expect to find life orbiting stars. Too little energy, and an organism cannot maintain itself or reproduce. There is not enough energy to do work. Too much, and an organism cannot hold itself together. The molecules bounce around too quickly to prevent undesirable reactions. As temperature increases, organisms would have to devote more and more of their resources to keeping everything in order and preventing unwanted reactions. In or near a star, energy would be abundant, but chemical reactions would rapidly break up any structure of complexity. In interstellar space, organisms would freeze. So, there must be some optimal distance between light and darkness, some distance from a star that makes life possible.

This does not mean that organisms could not exist nearer or farther, only that our best knowledge predicts life to be more likely in a certain zone. As we learn more about life and the universe, new insights could make stellar or interstellar life worth considering. Theorists will continue to explore the boundaries of biological chemistry in extreme environments. Experimental astrobiologists, however, have a more limited scope. We look at a hollow sphere known as the habitable zone. Not too hot, not too cold, but just right for life. This chapter will deal with the diversity, lifetime, and nature of stars in the known universe.[1]

Throughout this book, I treat astrobiology as the study of life as a planetary phenomenon. Some astrobiologists ponder life in other environments, and some interesting thinking has been done in those areas. Still, starting with Earth and intelligently searching the rest of space, I feel comfortable with the idea that life happens to planets. Life requires something to work with—a building material that can be shaped to store information and do work. Water, carbon, and other materials collect on planets and can maintain a useful distance from the star as the planet swings around in orbit. Does it have to be a planet proper? No, I think that any chunk of rock orbiting a star at a good distance would do. The question of what constitutes a planet has become tricky lately, after the discovery that Pluto shares the far outer Solar System with three other large objects. I only wish to say that a concentration of matter, whether planet, moon, comet, or asteroid, seems to be necessary for life as we know it.

Each of the next four chapters deals with questions about planets. Chapter 8 covers the formation, history, and properties of planets as they impact the existence—or possible existence—of life. Chapters 9 and 10 look at the particular properties of the objects orbiting our Sun. Chapter 11 covers what we know about planets in other stellar systems. In these five chapters, I hope to give you a better idea of the concrete location of life in the universe.

Making the Sun

If our story is to begin anywhere, it must be with the stars. I mentioned entropy and the tendency of everything in the universe to slowly slide toward disorder. Many people have wondered how life could have arisen at all in the face of this inexorable force. Indeed, for some, life seems to be evidence that such a force cannot be at work. How could something as complex as life happen in a universe slowly marching toward disorder? The first response involves star formation. Stars act as giant furnaces, radiating energy, assembling heavy molecules, and making little patches of space a great deal more comfortable. In short, stars make a small section of the vast coldness of space habitable.

Stars are, on a grand scale, evidence that entropy increases. Gravity acts to condense dust from a large swath of space into a single, bright sphere. During star formation, order and energy decrease over an immense volume of space. Meanwhile, order and energy increase within the tiny region around the star. If you have ever lit a really wide candle, you have seen a similar process. As the flame burns, a hole is eaten out of the melting wax, which slowly creeps up the wick only to be consumed, producing more heat. Another example would be crystal formation. If you put a seed crystal in the right kind of solution, crystal atoms will slowly accumulate, making the crystal grow. The solution mixes slowly and randomly, but, in the end, most of the atoms have become part of the crystal. Neither of these analogies really does justice to star formation, but perhaps they will give you some idea of how entropy can work to decrease order in a large area while order grows within a smaller area. Gravity and angular momentum act to collect vast amounts of hydrogen from space and form a star and planetary system.

Despite my attempt to include as much as possible regarding the history of life, we must pass quickly over the formation of the universe, a topic both interesting and controversial. Suffice it to say that, in all likelihood, there once was nothing, not even space to fill. The universe as we know it either came into existence as a result of random fluctuations in space-time, or perhaps was spawned from another, older universe. Cosmologists have a number of ideas, and I recommend looking into cosmology and Big Bang theory if you are interested.[2] However you choose to view the origin of space and time, it seems clear that the universe we know has been expanding since the beginning. Looking back in time, we can trace our universe's past for 13.7 billion years to the origin of history, a flash of light from hot, dense, and expanding gas. That flash is all around us as background radiation. From the details in that background and particle physics experiments, we can infer what happened even before that.

We have a good (if not complete) model extending back to the first seconds of the universe—but not before.

Sometime around 8 billion years later, our Sun was born. The local expanse of space probably contained very little mass from our perspective, perhaps 100 to 10,000 particles of dust and gas in every cubic centimeter. (By way of comparison, the air at sea level contains 10^{16} times that many particles.) The gas included hydrogen atoms (about three quarters), helium atoms (about one quarter), and trace amounts of heavier atoms. The dust particles—only one hundredth of all the mass—included chunks of matter up to 1 micrometer (one millionth of a meter) in diameter. These chunks formed through random interactions of gas particles in interstellar space.

So, the time is somewhere around 5 billion years ago and a diffuse gas fills up—or more accurately fails to fill up—local space. Local temperatures may have been in the tens and twenties Kelvin (K).[3] Some force manages to disturb the cloud and begins a process that will result in the birth of stars. Stellar winds or explosions in far distant stars send out waves of density in the gas and dust. Clumps of mass form as the waves move atoms and particles around and compress them. Once sufficient mass accumulates in one spot, the process starts a gravitational chain reaction with more and more mass joining the clumps. The immense size of molecular clouds (typically tens of light years across) and the random effects of clumping mean that hundreds of stars form throughout the cloud, with gravity wells sucking up giant spheres of gas and forming protostars.

Protostars contain nearly unimaginable amounts of mass settling together. (Our Sun currently has a mass of 1.989×10^{30} kg and had even more at its inception.) Over a period that seems long from a human perspective but incredibly short relative to the lifetime of a star, parts of the cloud collapse and start spinning. Most of the mass gathers in the center, while angular momentum builds up from random collisions. The cloud condenses into a protostar with a broad disk of material spinning around it. Within 10 million years, the protostar will collapse far enough to start nuclear fusion and a star will be born.

As the protostar collapses, atoms bump into each other more and more frequently, raising the temperature and starting chemical reactions. After only a few thousand years, the temperature reaches 3,000 K. The protostar starts to emit radiation. The radiation encounters infalling particles, increasing the temperature even more. Somewhere above 3 million K, the environment grows hot enough to support nuclear fusion. Hydrogen atoms crash into each other with enough force to stick and create a new helium atom. The process releases huge amounts of energy that radiate outward from the dense core of the star.

At some point, the Sun reached a rough equilibrium as hydrostatic and thermal forces came into balance. From then on, the Sun will change little over a period of billions of years. Hydrostatic equilibrium occurs as pressure in each layer evens out. At every point in the Sun, pressure from below and pressure from above match. Gravity pulls every point inward toward the center of the star, aided by the mass of all the points farther out. These forces are matched by pressure from the inside of the star as the points farther inward resist further compaction and try to expand as they heat up. The forces create a layered structure with the hottest and densest regions toward the center. Thermal equilibrium occurs as the temperature regime of the star stabilizes. At every point in the Sun, heat enters from the core and exits toward the surface at equal rates so that each layer has a roughly constant temperature. If the heat left too quickly, the star would rapidly cool and die. If the heat left too slowly, the star would burn up quickly and die. (Indeed, we will deal with both of these possibilities as we look at how stars die.)

Heat comes from the thermonuclear reactions in the core, and then gets released into space by radiative diffusion and convection. If a gas is transparent, heat can be transported by photons. Hence radiative diffusion means that radiation moves from atom to atom, traveling from hotter to

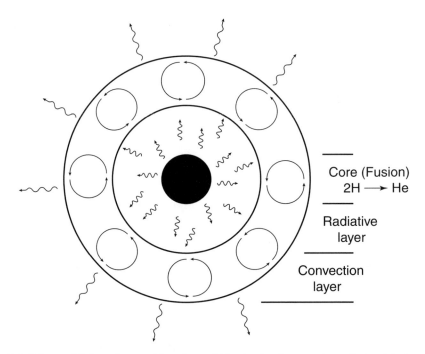

Core (Fusion)
2H ⟶ He

Radiative
layer

Convection
layer

7.1 Solar equilibrium diagram. Heat flow within the Sun happens by radiation (photons traveling through clear gas) and convection (circulation of atoms to exchange heat).

cooler regions. In opaque gases, however, which can be found nearer the surface of the Sun, heat must travel by other means. Convection occurs when a layer of material circulates; hot atoms move to cooler regions, where they can release their energy; cool atoms move to hotter regions, where they can be heated. In our Sun, the core produces heat, which needs to be released into space (Figure 7.1). A layer of ionized gas permits the radiation of heat outward through about 70 percent of the distance to the surface. An opaque gas layer sits above the ionized gas. Heat can only pass through this layer by convection. At the surface of the star, heat can once again be radiated. The majority of a star's lifetime is spent in such a state of equilibrium. Hydrostatic and thermal forces balanced, a star can persist with little change to its chemistry or size for billions of years, making helium out of hydrogen and releasing energy.

The Birth of Other Stars

Astronomers have deduced this history for our Sun based on observations of thousands of other stars all around the galaxy. Of course, no astronomer has been around long enough to see the whole process happen. They have, however, collected snapshots of molecular clouds, protostars, and stars at different moments in their life cycles. Remember that the universe is well behaved. We have every reason to expect that stars everywhere form in a similar manner. So, if we look at enough stellar formations in process, we can put together a scheme for how our own star came to be.

Our observations tell us more than that. Not all stars are small and yellow like the one nearest Earth. How do these stars form and how do they operate? The collapse of molecular clouds into disks, protostars, and eventually stars appears to occur similarly for all stars, but slight differences arise depending on specific conditions, especially mass and density. Molecular clouds vary in size and composition, so stars can form in a number of different colors and sizes. Different initial conditions lead to larger or smaller protostars. More massive stars burn hotter and redder.

The mass of a star also affects how it radiates heat. Bigger stars lose heat from their cores differently from our Sun. Stars more than four times as massive as the Sun have greater internal pressure due to gravity acting on all that mass. They also have an opaque layer next to the core. Heat has to escape through that layer by convection. At the same time, massive stars have a more diffuse outer layer farther away from the center of the star. The outer layer transmits heat by radiative diffusion. Thus, the larger stars lose heat in a fashion opposite to that of Sun-sized stars. The core produces heat. A dense layer around the core only allows heat outward by way of convection. Above the dense layer, heat radiates through a thin gas layer and out into space.

Stars smaller than our Sun have less complicated interiors. Below 0.8 solar masses, the internal temperature of the star never gets high enough to ionize an internal layer. The core produces heat and that heat travels all the way to the surface by convection before being radiated into space.

The Lifetime of the Sun

The lives of stars are measured in billions of years. Astronomers refer to the bulk of a star's lifetime as the main sequence. A star enters the main sequence when hydrogen fusion starts up in the core. A star leaves the main sequence when the fuel supply is depleted and core hydrogen fusion ceases. Between those two extremes, gradual changes take place as the star slowly consumes hydrogen. We measure a star's life by the chemistry taking place inside it.

The most dramatic changes occur in the core, where fusion takes place. The union of four hydrogen atoms produces a single helium atom, with the remaining mass being converted into energy and flung off into space. Every second our Sun burns, it converts 600 billion kilograms of hydrogen (matter) into pure energy. The energy (3.9×10^{26} joules) radiates into space, heating up the local environment and making life possible. The output of so much energy depletes the fuel reserve, slowly eating away at the mass of the star.

As the mass diminishes, so does the size. Weight from the outer layers pushes inward and the star maintains equilibrium by shrinking the core. Although the mass and volume of the core both drop, the volume drops more rapidly. Density and temperature increase. Hydrogen fusion speeds up, generating more energy and making the star brighter. Increased heat in the outer layers causes them to expand, increasing the overall size of the star. As a general rule stars in the main sequence can be said to change in this way: the core shrinks, condenses, and heats up; hydrogen fuses into helium faster; and the whole star expands and brightens.

In its 4.5 billion years on the main sequence, the Sun has grown slowly as it works its way through the internal supply of hydrogen: The radius has increased about 6 percent, the luminosity (or power output) has increased by 30–40 percent, and the surface temperature has increased by about 300 K. Incidentally, these changes pose certain difficulties for astrobiologists. The present-day Earth seems ideally suited for life, with solar radiation and terran atmosphere contributing to the stability of liquid water just about everywhere on the planet. If the Sun were cooler billions of years ago when life was first arising, what does that mean for liquid water? Astrobiologists call this the "faint young Sun problem." A cooler Sun 4 billion years ago would require a thicker atmosphere on Earth to keep the heat in and the water liquid.[4] The Sun will continue to become hotter and brighter. In perhaps 3 billion years, the Sun will put out so much energy that terran oceans will evaporate. Habitability, even on our own planet, seems to be deeply related to solar radiation and the evolution of stars.

Our Sun cannot burn forever. It must eventually run out of fuel, and, in about 7 billion years, things will start to fall apart in the interior. Once a star runs out of core hydrogen, it begins consuming the hydrogen immediately surrounding the core (the inner shell). The shell heats up, speeding up the rate at which the Sun becomes brighter. This emergency fuel reserve may extend the star's life by a few million years, but it also produces more helium. The helium, heavier than hydrogen, falls into the center of the core. The core continues to shrink and heat up while the outer layers of the star expand. The star becomes much more luminous. Surface temperatures start to fall, however, due to the enormous size and changing chemistry. At this stage, the Sun will be so large that Mercury, Venus, and possibly Earth will have been engulfed.

This too will pass. Helium filling up the core will eventually shut down hydrogen fusion altogether. Although the original source of energy has ceased, the temperature will quickly start to rise again. Remember that radiated heat from the core was one of the components of equilibrium, keeping the star stable during its main-sequence lifetime. When that force is no longer present, gravity collapses the core into an even smaller, denser sphere of helium. In the space of a few hundred million years—not that long for a star—the core in a Sun-sized star will shrink by about two-thirds (by radius). Gravitational energy gets converted into heat and the core temperature skyrockets. The temperature rises so high that new forms of fusion take place.

Giant Stars

At this point, the Sun will pass out of the main sequence, becoming a giant star. The core will be saturated with helium, stopping hydrogen fusion. In the surrounding inner shell, hydrogen fusion will continue to power the star, rapidly increasing the overall size and luminosity, but decreasing the surface temperature. Within a billion years after losing hydrogen fusion in the core, the Sun will have expanded a hundredfold and will be 2,000 times as bright as it is now. Despite the intense heat produced by the core and the amazing luminosity at the surface, the star will not produce enough energy to heat itself. The surface will cool and start to turn red.[5] The gravity of the star will also be insufficient to keep some of the gases in the outer layers from escaping into space. The Sun will start to lose significant mass.

Astronomers predict that the core of the Sun will shrink to a tiny sphere—about twice the size of Earth—made almost entirely of helium. When core hydrogen fusion stops, the core will not yet be hot enough to support helium fusion. Helium nuclei have twice the positive charge of hydrogen nuclei. The added charge means the electric repulsion (the force that keeps the north poles—or south poles—of two magnets apart) acts more strongly to separate the nuclei. Only extreme temperatures will speed up the helium nuclei enough that they will actually slam together and form heavier elements. As hydrogen fusion in the shells produces helium, the helium will rain down, making the core even denser and hotter. When the furnace reaches 100 million K, the core will start up again, burning helium this time, and producing carbon and oxygen nuclei.

Surprisingly, the process of helium fusion lowers a star's luminosity. Temperatures in the core will continue to rise. The core will start to expand, cooling the surrounding shells. Hydrogen fusion in the shells will slow down. Decreased hydrogen fusion results in a contraction and subsequent heating of the outer layers. If you find this confusing, you are not alone. Modeling stellar fusion involves an intricate balance of variable temperature, pressure, energy flux, and nuclear chemistry over a huge volume of space and a vast timescale. The immediate effect of helium fusion is a smaller, less luminous star with a hotter surface.

Main-sequence life depended on hydrogen fusion. While hydrogen was burning in the core, the Sun became bigger, brighter, and hotter. When that process ends, the Sun will expand rapidly and cool while core fusion is out of commission. Shell fusion, however, will continue, producing more helium. When helium levels rise high enough to ignite helium fusion, the process starts again at a higher level. Life after the main sequence will depend on helium fusion—at least for a little while. While helium burns in the core, the Sun will become smaller and less bright, but hotter.

Helium fusion accounts for a very small portion of a star's lifetime. Astronomers predict that our Sun's core will fuse hydrogen for around 12 billion years before switching fuels. After that, helium fusion can only last for 100 million years or so. At the end of core hydrogen burning, the star will expand to form a red giant. After helium burning kicks in, the star will shrink precipitously.

Once our Sun consumes all of the fusible helium in its core, the giant process begins again. An inert core of carbon and oxygen grows, surrounded by an innermost shell of helium fusion, then an inner shell of hydrogen fusion, and an outer shell. This time helium fusion is responsible for a rapid expansion. The star has less mass than the first red giant—hydrogen and helium have been lost into space—but it shines more brightly. Astronomers call stars in this stage of life "asymptotic giant branch stars" because their changes approach, but are not the same as, those of red giants. All fusion in the core and inner layers takes place in a tiny region about the size of Earth. Around this tiny region rests a giant envelope of diffuse hydrogen that still reaches to where Earth used to orbit. This outer envelope loses mass to space as radiation pushes particles outward and the star slowly dissolves. Asymptotic giants lose about a thousandth of a solar mass every year, losing mass more slowly than red giants but 10 million times faster than the Sun loses mass today.

As time passes, the star continues to lose mass. Helium fusion will stop repeatedly. Each time, hydrogen fusion will reignite the helium after a hiatus. When the helium reignites, it creates a bright flash of light that pushes even more matter away from the core, until nothing is left but an ember. The carbon and oxygen core slowly cools. Stellar processes will leave only an incredibly dense ball of matter drifting in space—a white dwarf. White dwarfs can pack a billion kilograms of matter into a cubic meter.

So the lifetime of the Sun will come full circle. The majority of mass, once a diffuse cloud of hydrogen and helium, becomes a fusion reactor for about 13 billion years before returning to a diffuse cloud. At the center, a super-dense cinder remains. The process is incredibly efficient, considering how large a volume of space it heats in that time.[6]

The Lives of Other Stars

Mass determines the life and death of stars much as it determines their birth. Stars with mass between 0.08 and 4 times the mass of our Sun (0.08–4 solar masses) appear to have lives similar to that of our parent star. Higher or lower masses result in different life histories. When a portion of a molecular cloud collapses, but has less mass than about 0.08 solar masses, a star cannot form. The low mass means low temperatures and hydrogen fusion cannot take place. The largest of these substellar objects possess enough heat to burn deuterium. They are referred to as brown

LIFE IN SPACE

dwarfs—too large for a planet, but not quite a star. Smaller objects simply burn out, forming cold spheres of matter. Stars with mass greater than 4 solar masses have shorter, brighter lifetimes than the Sun. Their mass allows them to fuse hydrogen and helium into many heavier elements through the course of their life. No one knows why stars cannot form above a certain mass, but no star has been observed larger than 130 solar masses and astronomers are confident that none will be.[7]

High-Mass Stars

In the cores of stars greater than 4 solar masses, the density and temperature reach extreme levels capable of producing heavier elements early in their life. Above 600 million K, carbon fuses into heavier elements. As the mass and temperature rise, more and more reactions can take place so that heavier and heavier atoms form. Above 8 solar masses fusion creates atoms as heavy as iron.

While on the main sequence, high-mass stars resemble their smaller cousins, though they have shorter lifetimes. When they exit the main sequence, however, they begin a far more dramatic process. The mass at the core of these stars is so great that fusion does not shut down after the formation of carbon and oxygen. The pressure continues to increase until the temperature exceeds 600 million K and carbon atoms begin to fuse. These dying stars go through the same expansion and contraction process as Sun-sized stars, but additional expansion and contraction occurs in which carbon fusion takes place, forming oxygen, neon, sodium, and magnesium.

A number of different reactions can take place, given sufficient mass. Each type of fusion means a more complex series of expansions and contractions during the death throes of the star. Hydrogen can fuse to form helium at a mere 40 million K, but for helium to end up as carbon and oxygen, a somewhat hotter temperature (200 million K) must be reached. Carbon, neon, and oxygen undergo fusion at progressively hotter and hotter temperatures. Finally, silicon tops the list at 2.7 billion K.

We can imagine each of these fusion types existing within a layer of the high-mass stars. Core hydrogen becomes depleted and the star undergoes expansion and contraction, igniting the core helium. A two-layer star results with a helium-burning core and a hydrogen-burning shell. The process continues in that fashion until a furnace layered like an onion burns in space, each layer raining down heavier and heavier elements into the core. Heavier-element fusion, occurring at extremely high temperatures, proceeds more and more quickly. In a star with 25 solar masses, astronomers expect hydrogen fusion to occur for around 7 million years. Helium fusion would occur for 700,000 years. By the time the star starts burning silicon, the core can be converted within a single day.

The process eventually stops with iron. All the reactions discussed so far produce energy. This energy comes from the strong nuclear force, which pulls protons and neutrons together within atomic nuclei. This force needs to balance the electric force, which pushes positively charged protons apart. Iron exists at the balance of the two forces—the formation of bigger nuclei means a loss of energy. In other words, fusion up to iron releases energy.[8] Fusion of iron traps energy.[9] Entropy wins again and solar fusion comes to an end. Stars with somewhat higher masses than our Sun—from 4 to 8 solar masses—go through a series of fusion stages, but eventually end up as our own Sun will. These stars lose most of their mass to local space, forming new nebulae. The remainder, a small compact sphere of heavier material, continues to smolder as a white dwarf.

Supernovae

Larger stars (greater than 8 solar masses) die in a far more spectacular fashion—they go supernova. The incredible mass consolidated in the stellar core at the time of iron formation begins a chain of events leading to the biggest explosions observed in the history of the universe. Fusion no longer holds back the force of gravity and the iron-rich core rapidly contracts and heats up. In less than a second, the temperature jumps to 5 billion K. The photons produced by this heated material are strong enough to rip iron nuclei apart and form numerous helium nuclei. Astronomers call the process photodisintegration. Millions of years are spent fusing hydrogen nuclei into heavier and heavier nuclei, and the process comes undone in an instant. The core is still contracting, however, and almost immediately a new type of chemistry starts. Electrons (negative) and protons (positive) have been forced so close together that they begin to react—despite their opposite charges. The reaction produces a neutron (no charge) and neutrino.[10] The neutrinos radiate out through the star, allowing the core to cool.

One quarter second after the final contraction begins, the core reaches a density similar to that of atomic nuclei.[11] Matter resists being packed any tighter than this. Core contraction stops abruptly. At the heart of the core, matter has already become too dense. It springs back outward in a process astronomers call "core bounce." The expansion of the core produces a tremendously powerful wave of energy that flies outward through the star. Meanwhile, the outer layers of the star have started to collapse inward. When core fusion stopped, they began to fall inward at amazing speeds.[12] The impact of the falling layers amplifies the wave produced by the core bounce. A shock wave occurs, traveling faster and faster as it passes through the less-dense outer regions of the star. When the wave reaches the star's surface, the star emits vast amounts of energy. A single explosion can release more energy than the output of our Sun over its entire lifetime. A supernova may momentarily become more luminous than billions of stars and outshine entire galaxies for a month.

The gamma rays emitted by a supernova can also have a significant impact on nearby planets. Some astrobiologists have suggested that the Late Ordovician extinction was caused by a supernova exploding within 6,000 light years of Earth.[13] Some event approximately 445 million years ago precipitated the death of 60 percent of all marine invertebrates, making it the second largest extinction event in Earth's history.[14] Astronomers suggest that a ten-second gamma-ray burst from a supernova could have radically changed the upper atmosphere, destroying the ozone layer. Such an event would leave surface and upper ocean organisms vulnerable to DNA-destroying radiation. By the time the ozone layer recovered, most species would be extinct. No direct evidence supports this theory, but astronomers believe that a supernova sufficiently large and sufficiently close to have this effect should occur about once every billion years. In other words, they expect it to have happened at least four times in Earth's history.

The Origin of Heavy Elements

Supernovae produce a number of astrobiologically significant effects. The shock wave can start star formation. The gamma-ray burst can affect nearby planets. And, most important, they produce and distribute heavy elements. As the shock wave passes through a dying star, it creates a wake of turbulent plasma. The wake forms pockets of immense density and heat where heavier elements once again form. Within these pockets fusion produces nuclei for gold, uranium, and all the other naturally occurring elements. The presence of such elements on Earth means that our own Solar System must have been formed from the remains of earlier supernovae.

Even carbon, the framework for life as we know it, represents the work of long-dead stars. Carbon does not require a supernova for formation, but neither would it be a common element in the universe after the Big Bang. Carbon must have been produced by helium fusion within the core of a star before being blown out into space. Carbon from dead stars would have been caught up in the initial protostar as our Sun formed and spun out into the planets. We owe our existence to the life and death of stars.

The Final Disposition of Giant Stars

After the explosion, high-mass stars pass on to one of two fates. The smallest of them collapse into neutron stars. Neutron stars represent huge mass (nearly 2 solar masses) compacted into a sphere roughly twenty kilometers across. They are so dense that unusual chemical reactions can take place in their interiors. Protons and electrons get squashed together to form neutrons. The name neutron star comes from this reaction. Larger high-mass stars have so much mass that they collapse into black holes. In black holes, so much mass has been compacted into such a small space that the force of gravity prevents the escape of everything, even light.

Solar Radiation and Energy

Astrobiologists are keenly interested in the radiative properties of stars. We think of the Sun like we think of Little Bear's bed in the Goldilocks story. It is not too hot and not too cold. It is just right for life because it emits light in the visible spectrum. Hotter stars produce more high-energy light. High-energy light has shorter wavelengths and tends to be bluer in color. Cooler stars and planets radiate heat and long-wavelength (lower-energy) light that tends to be redder in color than the Sun. Although the Sun appears white or yellow to us, it emits the most energy in green wavelengths (around 550 nm). We see white because of the relatively high abundance of all visible wavelengths.

The color of light turns out to be crucially important to life. Since most life on Earth depends on light as an energy source, the quality of that light has a big impact on the biosphere. Too much high-energy (short-wavelength or blue) light would be bad for organisms. The light would indiscriminately cause chemical reactions, breaking bonds and destroying organic structures. In fact, a number of organisms have developed surface pigments that absorb ultraviolet (UV) radiation. The surface protects internal biochemistry from the negative effects of intense radiation. The ozone layer on Earth also protects organisms from UV light that would otherwise destroy cells.[15] On the other hand, low-energy (long-wavelength or red) light would also be bad for life, because the photons would not be energetic enough to cause biologically interesting chemical reactions.[16] Our biosphere depends upon abundant solar energy to string carbons together and do work. So life seems to be intimately related to white light in what we think of as the visible spectrum.[17]

This should not come as a surprise. Much as we find Earth ideally suited for life, so we find "visible" light to be ideal. Visible light is visible precisely because it has the right wavelengths to effect chemical reactions in the human eye—we prefer it because of its chemical properties. Light in the range of 400 to 700 nm can be used to do chemical work, both in the human eye and in the phototrophic reaction centers of bacteria. Life would not function any other way. In other words, we have evolved the ability to capture the available photons and use them to do work. The quality of light determines the kind of reactions possible and the type of life that will arise.

So we believe that life around "moderate" stars would be more likely than life around hot blue stars or cold red stars. Blue stars produce too much disruptive ultraviolet radiation, and red stars produce too little chemically useful visible light. This is, of course, a huge generalization. Our Sun provides far more energy for the terrean biosphere than can be used by life as it currently exists. Hotter and cooler stars could still produce enough radiation at useful wavelengths. At the same time, we recognize that white stars appear to be more ideally suited for the evolution and survival of life in space.

8

The Planetary Phenomenon

Life requires a very special place in the universe. Water, carbon, and energy constitute the basics and they must be abundant. In this chapter, I turn to the specific locations where we can find these commodities. The vast majority of the cosmos consists of a cold void with a few atoms of hydrogen and helium wandering around; we can say with some certainty that life (as we know it) would not be found there. No, life needs a warmer place with a little more furniture. Stars provide a universal heating system. They not only produce energy to heat up local space, they come with their own collection of solid matter in the form of planets. These packages of solid elements include iron and silicon as well as the hydrogen, carbon, nitrogen, oxygen, phosphorus, and sulfur we have come to associate with life. The combined elements come so closely packed together in the stellar neighborhood that they start to react with one another. Most notably, we see fusion at the core of stars, but other interesting chemistries happen a little farther out. Processes such as life require available matter and a stable platform on which to live.

What happens to all the material left over in star formation? Not all of the matter from collapsing molecular clouds goes into creating a star. Just as your arms fly outward when you spin quickly in place—just as a centrifuge forces matter outward by spinning rapidly—so the angular momentum in protostars throws out small amounts of matter, forming a spinning disk. The center of this disk collapses into a star, but the remainder stays in orbit, forming a protoplanetary disk. Planetary scientists debate the mechanisms by which the disk collapses into more recognizable bodies, but they do know that it forms a preliminary stage on the way to planets, moons, asteroids, comets, and all the other debris found floating around stars.

I mentioned earlier that we have only one example of life, and that severely limits our ability to investigate the phenomenon of life. One example does not allow for comparison, contrast, or any other kind of complicated observation. Until recently, planetary scientists faced the same problem. How do we figure out the formation and nature of planets when we can only see one stellar system—our own? How do we know which properties are essential and which are rare anomalies? How do we know how the whole thing works unless we can compare it to something?

In the last two decades, astronomers started to address this problem in a serious way. Through various methods, including both inference and direct observation, we now know of over three hundred planetary objects orbiting other stars. These objects present an amazing opportunity to expand our understanding of planet formation. Recent theories show a radically different understanding than that which went before. Because of this, ideas about planet formation remain in flux. In several areas, such as large planet formation, more than one theory exists; the rapid accumulation of new data means that we are constantly adjusting and gradually coming closer to a full understanding of a very complex system.

The Protoplanetary Disk

In a general sense, planet formation resembles star formation at a smaller scale.[1] A diffuse cloud of dust and gas condenses as a result of gravity and angular momentum. The condensed matter forms a spinning sphere in space. One major difference, of course, comes from the size of the sphere. The immense quantities of matter present in a star result in tremendous pressure at the core. The pressure causes the core to heat up and start fusion. Planets, being smaller than stars, do not have sufficient mass and heat to burn hydrogen. Some planets may have molten cores, where iron and other elements exist in a liquid form. That can only happen, however, when sufficient quantities of matter accumulate in the right quantities. For the most part, objects orbiting stars will be solid rock or ice, along with lighter elements such as hydrogen and helium in gaseous form.

The protoplanetary disk forms from leftover matter that has not condensed in the formation of a star. Current thinking holds that some stars produce these disks in their formation, whereas others do not. Because so few planetary systems have been observed and because of the huge number of stars out there, we cannot know yet what proportion of stars produce this ring of additional matter. The rotation developed as the star collapses from a nebula gets passed on, so that the disk and the star end up spinning in the same direction. Protoplanetary disks are much denser than nebulae, but still pretty sparse. They contain mostly hydrogen and helium, with varying amounts of heavier elements produced in earlier stars.

Current ideas on planet formation divide planets into two categories. The first category—including Earth, Mercury, Venus, and Mars—represents planets made up of mostly rock with a thin layer of atmosphere on the surface. These planets have traditionally been called "terrestrial" because of their similarity to Earth. Astrobiologists, myself included, try to refer to them as rocky planets. This works to eliminate an Earth-centric bias and to avoid confusion when talking with biologists and geologists for whom "terrestrial" has a different meaning. A second category, called giant planets or gas giants, includes Jupiter, Saturn, Uranus, and Neptune. These planets tend to be much larger than Earth. They may have a relatively small, rocky core, but the majority of their mass and size comes from their gaseous atmosphere. The term "gas" here is somewhat of an oversimplification. The mass of giant planets presses inward, compressing much of the hydrogen into a liquid ocean surrounding a rocky core.

The Planets

For both rocky planets and giant planets, the core appears to be a collection of solid minerals heavy in iron and silicon. Atmospheres may be less consistent. The largest giants in our own system, Jupiter and Saturn, contain predominantly hydrogen and helium. This comes as no surprise, considering how abundant those elements are and how much mass is needed to form so large a planet. The atmospheres of smaller planets show greater diversity. Uranus and Neptune possess large quantities of methane (CH_4) and ammonia (NH_3), whereas the rocky planets have more carbon dioxide (CO_2) and pure nitrogen (N_2). Earth is unique among local planets in its high levels of pure oxygen (O_2 and O_3). This seems to be a result of life and will be discussed later when I describe the history of life on Earth. Abiotic processes have a tendency to reduce atmospheres as the oxygen becomes associated with rocks or lost to space.

Looking at our own Solar System, we see rocky planets in the inner system and gas giants in the outer system. Planetary scientists still do not know if this pattern holds for other stars. Current methods for detecting extrasolar planets favor larger planets and bias our sample. Rocky planets may arise closer to their star due to the higher concentration of heavy elements. Alternatively, gravitational effects from the Sun might make it more difficult for large planets to form closer in. Planetary scientists once favored the latter theory, but the discovery of many "hot Jupiters"—huge planets orbiting close to their parent star—makes it seem less likely. It would be a mistake to assume that other systems behave the way that ours does without further evidence.

I should say a word here about the abundance of rock close to the Sun and the abundance of ice farther out. Astronomers divide substances into two categories—refractory and volatile. Refractory components of the protoplanetary disk, such as iron and silicon, have very high melting points. Thus, they exist as solids throughout the system and are free to interact and cling together everywhere. Volatiles, on the other hand, such as water and methane, have low melting points. Closer to the Sun, they heat up and form gases. As gas, they do not cling together and avoid getting trapped in planet formation. In the outer system, temperatures are low enough to freeze even volatiles, and we see water-ice abundantly.

Formation of Rocky Planets

Present models suggest a period of around 30 million years for the gradual accumulation of matter into rocky planets. (This would have occurred between 4.6 and 4.5 billion years ago for our system.) Growing chunks of rock slowly clear out the space around the star as bits of mass collide and stick together.[2] The process begins as tiny dust particles in the protoplanetary disk settle into the midplane and begin to interact. The dust particles, interacting with one another, form larger grains as they race around the central star. The grains encounter still other grains and stick together, but they may also drift together in clusters that undergo gravitational collapse. The cluster suddenly achieves sufficient mass to fall together and forms a larger rock all at once. At this stage, planetary scientists refer to the growing rocks as planetesimals, or tiny planets. The method by which grains grow to planetesimals remains somewhat mysterious because gravity may not be sufficient to gather pieces under a kilometer in diameter. Turbulence and chemical attraction may play an important role.

The planetesimals remain subject to the same forces as the smaller grains, initiating "oligarchic growth." In politics, oligarchy refers to rule by a small number of powerful individuals. In planetary science, it reflects the gradual accretion of mass into fewer and fewer individual planetesimals until a small number dominate the system. As they decrease in number, they increase in mass and size. More mass means more gravity. More size means more surface area to interact with other planetesimals and tiny grains of rock. The large planetesimals clear out the remains of the protoplanetary disk in the inner system. Like giant vacuums cleaning up local space, they zoom around the parent star until most of the matter is contained in perhaps 100 large planetesimals or "planetary embryos." At this stage, the embryos have about as much mass as Earth's Moon.

Over millions of years, the planetary embryos collide, reducing their number even further. Gravitational interactions affect their orbits, with larger embryos (and planets in the outer system) pulling the smaller embryos along or gobbling them up. Eventually, only a few planets remain in clearly defined orbits. These orbits ensure that the planets neither drift outward under the influence of other planets and momentum, nor drift inward under the influence of the star's gravity. Our own stellar system seems to have achieved equilibrium, with the planets and countless smaller pieces of matter orbiting in fixed paths for billions of years. No one knows how common this is in the universe, or even in our own galaxy. Planetary scientists currently think that equilibrium should be easy to achieve.

Formation of Giant Planets

Giant planets form somewhat differently than rocky planets. Observations of protoplanetary disks around other stars suggest that the disks only last for 1 to 10 million years. Because so much of the matter in the disk would be necessary, giant planets are thought to form within that time frame. Planetary scientists currently favor two possible theories for giant planet formation, accretion and collapse.[3]

In the accretion model, giant planets start their lives much as rocky planets do. They result from the long-term interactions of particles and grow from chunks to planetesimals, planetary embryos, and eventually planets. Giant planets acquire much larger quantities of volatiles than the rocky planets, but the process is much the same. Problems with this model come from the time scales necessary. Why would the inner planets take so much longer?

In the collapse model, local, randomly formed clumps of matter in the protoplanetary disk cause gravitational instabilities. Dust and gas from a wide region of space suddenly collapse to form the giant planet, with the heavier elements settling to the middle of the new sphere. The new planet remains caught by the gravity of the parent star and can continue to accrete more mass as it sweeps around its orbit. Farther out, planets grow larger because volatiles can freeze and contribute to the mass of planetary cores. The larger cores and colder gases allow these planets to attract a very deep atmosphere.

Models of giant planet formation need to account for migration as well as origin. Planetary scientists suspect that a number of extrasolar planets started as giant planets far from their parent star before moving inward for a closer orbit. Interactions among objects within our own Solar System are also consistent with giant planet migration. Unfortunately, current observations and mathematical models seem to be consistent with migration both inward and outward, and we lack the data necessary to understand giant planet movement well. Extrasolar planet research promises new insights in the coming decades.

Life Beyond Planets and Moons

Science-fiction writers and a few astrobiologists have suggested numerous alternatives to planetary life. Life could occur anywhere, from really hot life on the surface of stars to really cold life distributed among interstellar dust clouds. We might even conceive of organisms that fly between the stars, using saved-up energy. Postulating strange forms of life can be interesting and particularly useful as we explore the meaning of life. Unfortunately, this broad approach to the definition of life has two drawbacks in practice: practicality for observation and practicality for existence.

Life requires both energy and matter. Life too close to a star would be short-lived because it could not prevent unwanted reactions. Life too far away would be short-lived because it would not have the energy to maintain itself.

Life that does not occur on a planet orbiting a star may also be extremely difficult to observe. How would we find life existing in open space, given the vastness of territory to search? How would we find life within or on a star, given the immense amounts of radiation emitted? Perhaps these will be reasonable pursuits in the future, but at present, planets will remain the best place to look for life.

The Habitable Zone

The question of where life might be feasible can be answered in a number of ways, depending upon your definition of life and what limits you place on organisms. For the practical purposes of astrobiologists, the habitable zone describes the area within a stellar system in which life might exist.

Chapter 6 introduced three basic requirements for life in the cosmos— water, carbon, and energy. Each one occurs abundantly in stellar systems, but they do not all occur in the same places. In looking for a region of habitability, the three need to be balanced. The requirements fit well with the Goldilocks principle: We want a planet (or moon) that is neither too hard nor too soft, but something just right. We want a planet far enough from the central star for water to be present. Volatiles tend to be more common in the outer solar system, where the temperature is lower. Water in the inner system does not accrete to planets as easily, and when it does, it can be lost. Surface and atmospheric water may evaporate at high temperatures and be blown away by stellar radiation (as happened on Mercury).

On the other hand, the planet needs to be near enough to the star for heavy elements including carbon to be abundant. Life requires material to work with. As you move farther and farther out into space, volatiles increase and the relative abundance of carbon decreases. In our own Solar System, planets beyond Mars consist mostly of volatiles like water and methane. Heavy elements are harder to come by. Thus, we want a planet to be close enough to the central star that heavy elements are available, but far enough for water to be present.

Some planetary scientists see a problem with the two criteria mentioned so far.[4] They point out that only the inner planets have enough carbon, and only the outer planets have water. The inner system may be too hot for large chunks of ice to form during the planetesimal stage of planet growth. Close to the star, large quantities of ice will simply evaporate. How then did water get from the outer Solar System to the surface of Earth? The best answer currently proposed requires that the rocky planets form within the inner region first and then acquire water later. Comets formed in the outer solar system develop irregular orbits that swing them into the inner region, where they collide with planets. Impacts may also have had an important role in delivering organic carbon to the surface of Earth. Abiotic processes in the outer system create simple organic compounds like amino acids.

Finally, we must think about energy concerns. The star acts as both a power source and a heater for our region of habitable space. Organisms need to be close enough to the star to take advantage of the abundant energy, far enough from the star to be safe from unwanted reactions.

Looking at these three criteria, we are left with a habitable zone extending around the star in a giant donut. Planets, the platforms of life, occupy a disk around the star; they developed from the planetary disk. The habitable zone should be in the plane of that disk. The water line defines the inside border. Inside that line, water evaporates on the surface of planets, if it exists at all. The ice line defines the outside border. Beyond that line, water on the surface of a planet will freeze, stopping biochemistry and inhibiting life. The ice line might be extended by planetary atmospheres that keep the temperature in, but even then there is a limit. Biospheres need the energy to keep running. Figure 8.1 shows one conception of the habitable zone for main-sequence stars.[5]

Nonstellar energy sources will also affect the outer boundary of the habitable zone. Giant planets like Jupiter can provide additional energy to their moons in the form of tidal and magnetic energy (see jovian satellites in Chapter 10). Smaller habitable zones, then, can occur around giant planets.

Figure 8.1 does not take into account the amount of time a planet

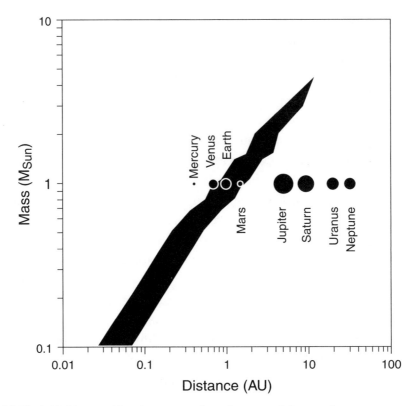

8.1 The habitable zone. Here it corresponds to the range of distances from a parent star amenable to planets with liquid water. The model takes into account atmospheric conditions, and thus depends on a number of variables. See note 5 for details.

would remain within a habitable zone. Stars change throughout their lifetimes and planets can shift in their orbits. Consistency is incredibly important to our concepts of life and habitability. Although our discussion covers planetary distances, the habitable zone still occupies only a narrow range of space around each star. A habitable planet or moon may need to stay within that zone for millions of years to foster the origin of life. This means that habitable planets need very stable orbits. The Solar System contains planets with extremely well-behaved orbits, but many stellar systems contain planets with more dramatic orbits, passing in and out of the habitable zone twice every year. A planet may be a very nice place to live for a million years, and then pass out of the habitable zone.[6] Luckily, Earth appears to have a very stable orbit. It will most likely stay within the habitable zone until our star explodes in about 7 billion years.

Planets act as platforms for the development of life in space. Carbon and other heavy elements are concentrated in a local area, providing abundant reactants and long-term stability. Planetary scientists use planets in our own Solar System as well as extrasolar planets to try to understand the processes that shape these important bodies.

Planets form as leftover matter from star formation accretes into giant hunks of rock and gas orbiting the new star. Rocky planets, including Earth, contain mostly refractory materials like iron and silicon, with a narrow gaseous atmosphere. Gas giants, such as Jupiter, form from mostly volatile gases like hydrogen and helium, with a small solid core in the center. Astrobiologists continue to puzzle over the processes that allowed Earth, a rocky, inner planet, to acquire so much water—a volatile much more common in the outer Solar System. As we take a closer look at our local planets, bear in mind the importance of the habitable zone. Where might we find carbon, water, and energy in the right quantities to support life?

9

The Inner Solar System

We now turn our attention to our own Solar System, the set of objects orbiting the local star sometimes called "Sol," but usually referred to as the Sun. It can be divided into five regions. The first three should be well known to most readers while the last two may be somewhat unfamiliar. The first and inmost region includes four rocky planets—Mercury, Venus, Earth, and Mars. Just outside of this, a giant ring of rocks circles the Sun, the asteroid belt. Beyond the asteroid belt lie four giant planets—Jupiter, Saturn, Uranus, and Neptune—with their moons and rings. Past Neptune we find a second ring of loose rock—the Kuiper belt. Among the largest Kuiper belt objects (KBOs) is Pluto, until very recently considered a planet by astronomers. Beyond the Kuiper belt lies a vast store of icy chunks known as the Oort cloud. Whereas everything within the first four regions lies neatly in a single plane, the Oort cloud forms a giant sphere around the Sun, stretching in every direction.[1]

Orbits happen for a very simple reason—gravity. As a stellar system forms, planets come into being with a great deal of momentum. Remember that the protoplanetary disk started off spinning. The conservation of angular momentum ensures that the spin of the disk survives in the orbit of the planets. Each object jets around the Sun. It constantly falls in toward the Sun—because of gravity—but always misses—because of momentum. Imagine a ball on a rope. If you hold onto the rope and spin fast enough, the ball will fly around you. The string pulls it toward you while momentum keeps it moving. Planets act like the ball, settling into stable orbits where the two forces equal out.

Perhaps you are accustomed to the terminology of stars and planets, but physicists have a far more generic language for orbiting objects. They call the orbiting object a satellite and the orbited object the parent. Thus Earth is a satellite of the Sun, and the Moon is a satellite of Earth. Satellites rarely travel in perfect circles, so the relevant number here is the semimajor axis. It measures the average distance, calculated by adding the longest distance—when the satellite is farthest from the parent—to the shortest distance—when the satellite is closest—and dividing by two. Planetary scientists often report these distances in astronomical units (AU). One AU corresponds to 149.6 billion meters (1.496×10^8 km), the semimajor axis of Earth's orbit.

Three other numbers give us a very clear idea of how a satellite behaves. First, eccentricity describes the roundness of an orbit. A perfectly circular orbit has an eccentricity of zero. Objects with too much momentum to stay in orbit have eccentricity of one or greater. Thus, all satellites have eccentricities that are positive but less than one. The larger the number, the more irregular the orbit. Orbits in the Solar System are amazingly regular (possessing low eccentricity). Second, inclination describes how far an orbit strays from the ecliptic—the plane defined by Earth as it passes around the Sun. The larger the angle, the greater possible separation between a planet and the plane of reference. Solar planets vary from the ecliptic only slightly. (For moons, the planetary equator replaces the ecliptic as the reference plane.) Third, obliquity describes the degree to which a planet tilts. Earth, for instance, spins a little over twenty-three degrees to one side. In other words, the axis about which we rotate (Earth spinning) makes a twenty-three-degree angle with the axis about which we revolve (Earth orbiting the Sun).

Here begins my quick tour of the Solar System for amateur astrobiologists. Chapter 10 covers the giant planets, Kuiper belt, and Oort cloud.[2] In this chapter, I explore the rocky bodies that dominate the inner Solar System. Four planets and the asteroid belt travel through this region of space. They have high concentrations of iron and silicon and little to no atmosphere. Volatile materials like water and methane occur rarely, with Earth as the one exception.

Mercury

Mercury comes closer to the Sun than any other planet. At a little over one-third the distance of Earth (0.39 AU) from the Sun, the heat on Mercury can be intense. Daytime temperatures reach 800°F, but nighttime is a different story. Mercury has virtually no atmosphere, meaning that nothing holds the heat in during the night. The dark face of the planet reaches −280°F. The extreme temperature shifts preclude liquid water and make Mercury a very unlikely place to find life.

Planetary scientists think that a giant impact early in Mercury's history rearranged the composition of the planet, ejecting both atmosphere and some quantity of lighter elements in the rock. Most of the planet appears to be composed of iron, making it the second densest planet in the system—almost as dense as Earth. The surface of Mercury is solid, prohibiting the formation of volcanoes. A very thin atmosphere results from transient particles and gases blown off the surface by meteor impacts and the solar wind. Earth-based radar astronomy suggests the presence of water ice deep within craters near the poles.

Many factors make Mercury an excellent place to look for the history of the Solar System. Craters from meteor impacts can be used to investigate the amount of stray matter flying around the inner Solar System at various times. Craters form on all of the rocky planets, but only Mercury keeps them safe from erosion, which tends to smooth away such features over time. Venus, Earth, and Mars all show some volcanism, with magma spreading out and resurfacing the planet. They have winds that blow dust over the craters. Water rushes over the surface of Earth craters and changes the landscape. Only on Mercury can we see the pattern formed over the entire history of the system.

Exploration of Mercury has been minimal. NASA's *Mariner 10* spacecraft visited the planet in 1974, taking pictures of about 45 percent of the planet's surface. The *MESSENGER* spacecraft (also NASA) left Earth in 2004 and is expected to achieve orbit in 2011. The European Space Agency and the Japan Aerospace Exploration Agency (ESA and JAXA) are currently planning a mission called *BepiColombo* for launch in 2009.

Venus

Venus presents a much more interesting target for exploration. The planet orbits the Sun at 0.72 AU, making it similar to Earth in several ways. The thick atmosphere and volcanic activity allow for many exciting kinds of chemistry. Venus closely resembles Earth in a number of ways. Both planets are roughly the same size. Venus has a nearly circular orbit and a heavy atmosphere. It may even have tectonic activity with surface plates shifting across the mantle to make continents.

Unfortunately, the surface environment is hostile to spacecraft and all forms of life as we know it. The violence of venusian environments comes from the incredible heat and toxicity of the atmosphere. The major component is carbon dioxide, a greenhouse gas. Heat from the Sun hits the planet and gets trapped in the atmosphere instead of radiating back out into space. Surface temperatures reach 880°F, exceeding even Mercury. Photochemical processes in the upper atmosphere produce sulfuric acid (H_2SO_4) rain. Sulfuric acid is an extremely corrosive acid used to etch glass and dissolve rust off the surface of steel. This acid rain would easily destroy carbon-based structures on the surface, even if they could develop. To make matters worse, the venusian day (period of rotation, 243 Earth days) exceeds the venusian year (225 Earth days). This means that the atmosphere on one side of the planet regularly heats up and creates hurricane-force winds.

Another problem for life on the surface of Venus comes from frequent magma flows. Liquid rock from the interior of the planet regularly flows out onto the surface. A major volcanic event must have occurred around 500–300 million years ago. While multicellular life was taking over on Earth, liquid basalt was moving over 90 percent of the venusian surface.

Both NASA and the Soviet space agency devoted themselves to understanding Venus in the last half of the twentieth century. Its size, orbit, and atmosphere made it seem a good candidate for harboring life before we discovered the violence at the surface. Twenty-two missions have returned data, including *Mariner, Venera, Pioneer Venus, Vega, Magellan,* and *Galileo.* Orbiters and atmospheric probes produced most of the data; landers never survived long under the intense heat and pressure. The longest contact with a lander, one of the *Venera* spacecraft, lasted only 127 minutes. The missions did produce maps of 98 percent of the surface and a good understanding of atmospheric chemistry and dynamics.

Conditions on Venus have led planetary scientists to drop it from the list of possible locations for life. Mars seems a better candidate and recent missions show a much greater emphasis on exploring the Red Planet. ESA and JAXA have both planned for missions to Venus within the next two decades, *Venus Express* and *Planet-C* respectively.

Earth

From our perspective, Earth is a pretty special place. A number of science and science-fiction authors have speculated on why life may have arisen here, but at present we still do not know. Silicon and iron predominate in the rocky portion of the planet, much like the other rocky planets. The mass and density of Earth facilitate tectonic activity, with regular cycling of material in and out of the crust as well as constant rearrangement of the surface. In addition, Earth has large reservoirs of water held in the crust and present in large liquid oceans.

Plate tectonics impacts terran life significantly. The gradual movement of Earth's crust ensures the constant cycling of carbon, nitrogen, sulfur, and other biologically important elements. Undersea vents and volcanoes provide dynamic environments where reduced compounds—with available high-energy electrons—come to the surface. Geologists remain uncertain about the exact conditions that led to Earth's dynamic crust, but comparison with other planets demonstrates the immense importance of this kind of activity. Mars, possibly because of its small size, never had enough internal energy to support plate tectonics. Its period of geologic activity ended millennia ago. Venus, on the other hand, has plenty of internal heat, but cannot release it gradually. On Venus, pressure and heat build up inside the planet until a giant eruption of magma breaks through the crust and completely resurfaces the planet. Earth benefits from an unusually high amount of water in the crust, which lowers the viscosity and facilitates heat transfer from the core.

Earth, unlike many planets, is geologically active. Materials from inside the planet get cycled around near the surface and often break free. A layer of convective rock lies miles below the surface. Plates of solid rock float on top like sheets of ice on the surface of a pond. Magma flows up from spreading centers (where the plates move apart) and volcanoes at subduction zones (where one plate slips under another). Rock from the surface also gets sucked under at subduction zones. Both processes guarantee that minerals such as nitrogen and reduced metals get cycled through the atmosphere and geosphere. Tectonics acts as a giant circulating system, making sure elements and electrons never get trapped in one place but constantly move about.

The characteristics of Earth seem amazingly suited to circulation and mixing of materials. Ground, sea, and sky all have their own cycles. Large oceans, which cover 70 percent of the surface, slosh around the planet, drawn by the pull of the moon. These tides result in regular mixing within the oceans and dynamic interfaces between sea and land.

The atmosphere contains mostly nitrogen (N_2, 78 percent) and oxygen (O_2 and O_3, 21 percent), along with varying amounts of gaseous water. Notably, the oxygen has not always been there, but appears to be the result of a biological revolution sometime between 2 and 3 billion years ago. The oxygen produced by plants and other photosynthetic organisms acts as a shield for ultraviolet (UV) radiation. I mentioned in Chapter 7 that one problem for life was an overabundance of energy. We have seen how too much heat (long-wavelength radiation) can make liquid water impossible and life unfeasible on Venus. Another consideration comes from UV (short-wavelength) radiation itself. UV rays reaching the surface would have a tendency to break bonds and scramble organisms. Nucleic acids—life's mechanism for storing information—are particularly susceptible to UV. Oxygen in Earth's atmosphere plays an important role in the uniqueness of Earth by protecting organisms at the surface.

Temperature on Earth can be complicated. Overall, the temperature remains fairly constant, thanks to Earth's nearly circular orbit and atmospheric buffering. The atmosphere retains some heat from the Sun—only a small percentage of the heat retained on Venus, but enough to ensure that nighttime temperatures do not drop too low. Earth tilts relative to the plane of the ecliptic so that the northern and southern hemispheres get different amounts of solar heat at different times. Earth currently has an obliquity of 23.45°. The difference in temperature results in both seasons and some weather effects because heated portions of the atmosphere behave differently from unheated portions. Mercury and Venus have almost no tilt—2° and 2.7° respectively—whereas Earth and Mars are nearly the same in this regard. Recorded temperatures on Earth range roughly from −130° to 130°F. Almost half of this range accommodates liquid water, making Earth a good place for life overall. With the extreme negative temperatures only present near the poles, water stays liquid year-round over the majority of the surface.

The study of greenhouse gases has received a great deal of public attention in recent decades, so I feel I should say a word or two about the subject. The carbon dioxide in Earth's atmosphere plays an important role in the regulation of temperature. The process of "greenhouse warming" occurs when gases in the atmosphere reflect energy back to the face of the planet. When solar energy comes in from the Sun, the planetary surface sends energy back as heat. Greenhouse gases are those gases that reflect that heat back toward the surface.[3] Earth's greenhouse situation is not as bad as Venus's, nor could it ever be. On the other side, Mars does not have a thick enough atmosphere. Temperatures there stay below freezing year round, despite the fact that it is nearly close enough to the Sun to support liquid water—given the right atmosphere. Mars needs more atmosphere and more greenhouse gases. Venus needs less.

The question of greenhouse warming cannot be simplified to the simple statement that greenhouse gases are bad. The situation remains quite complex because Earth's atmosphere keeps the temperature constant through a number of mechanisms. Greenhouse gases are one of the more important ones. Carbon dioxide from nonhuman sources provides just the right amount of warming to keep the planet at the temperature we want it. Most atmospheric scientists fear that if we mess with the thermostat too much, the whole system will explode in our faces, resulting in wild, unpredictable weather. What is worse, we know that there is a strong correspondence between surface temperature and carbon dioxide in the atmosphere. More CO_2 means higher temperatures. The mechanism remains somewhat mysterious, though the model of a greenhouse works pretty well. We still have trouble understanding how the system deals with excess CO_2 when it has to. It could be that natural processes sequester it in the oceans. It could also be that the oceans are already saturated. However the mechanism works, though, it remains clear that carbon dioxide emissions to the atmosphere work like a thermostat. More emissions and the temperature goes up. Fewer emissions and the temperature stays the same.

Much has been made of the distance between Earth and Sun as it relates to temperature and radiation. The habitable zone has often been defined solely in terms of temperature necessary for liquid water. An additional energy concern arises, however, from the energy flux. Huge quantities of energy reach Earth as photons. Greater than 99.9 percent of the energy available on Earth comes from solar radiation, amounting to approximately 170 petawatts or 1.7×10^{17} joules per second. Solar radiation at Earth peaks in the visible range (roughly 400 to 700 nm wavelengths), ideal for breaking and forming many chemical bonds.

Energy from the star has been trapped within the planet's biosphere. A large number of complex biological systems capture solar energy and store it chemically and structurally. Life has colonized every available portion of the surface in an amazing range of conditions. Life extends from burrowing individuals miles below the surface to tiny floating individuals high above. These organisms all appear to be intimately related. Many organisms exist in remote or challenging areas (such as the sea floor or desert) where water, carbon, or energy may be difficult to find. Earth benefits from a unique combination of constancy (e.g., temperature) and change (e.g., geologic activity). Which factors were necessary for the origin of life? We still do not know. Perhaps all of these factors must be just right for life to occur; perhaps only one.

Research on Earth has been extensive, and it would be impossible to list even the relevant NASA missions. Many nations have launched satellites returning important data about the surface and atmosphere. Weather prediction remains a top priority. Study of the age and composition of rocks and soil also plays an important role in choosing farming sites as well as mining and extraction of oil and natural gas. These studies would continue even if pure science research—including astrobiology—were to cease. Astrobiologists, however, enjoy the unique perspective of putting the different data together to form a coherent whole. Astrobiologists get the chance to see Earth not just as home but as a living planet in space. Such a perspective will be essential to understanding Earth as a single, interconnected system.

Chapters 12 through 19 cover Earth and terran life in much greater depth. Life, after all, sets us apart from all other objects in the Solar System. If life does exist elsewhere, it is nowhere near as visible. Astrobiologists attempt to connect the dots between planetary science, chemistry, and biology to see why this seemingly unique condition developed where it did.

The Moon

Of the four rocky planets, only Earth and Mars have moons and only Earth has one large enough to be spherical. The moons of Mars are quite small, irregular chunks. The Moon, on the other hand, has about a hundredth the mass of Earth, making it five times as massive as Pluto. The Moon's small size and almost nonexistent atmosphere mean that it would be a poor candidate for finding life. Much like Mercury, however, it can tell us a great deal about the history of the Solar System. The lack of atmosphere and geologic activity means that it preserves a record of impacts all the way back to its formation.

The formation of the Moon has been an area of great interest for planetary scientists. It's an irregularity in the inner Solar System, having a similar composition to the four rocky planets, but with far less iron. Research from the *Apollo* missions supports the idea that the Moon formed soon after the origin of Earth. A Mars-sized planetary embryo collided with the planet and vaporized the outer layers of both bodies. After collision, Earth reformed, containing the dense core of both objects. Lighter elements piled in around the core or entered a close orbit. The orbital debris eventually coalesced into a moon, much the way the planets coalesced, but on a smaller scale. This scheme explains both the high density of Earth and the low density of the Moon relative to other rocky planets.

The Moon has an interesting, and as yet poorly understood, impact on terrean life. It causes tides within the surface water of Earth, creating another type of mixing of chemicals on the surface. It may also have some effect on the stability of our movement through space. Astrobiologists remain uncertain of how important these features are.

Mars

Astrobiologists see Mars as the best hope for finding nonterrean life within the Solar System.[4] Mars travels around the Sun about half again as far as Earth (1.52 AU). It resembles Earth in a number of important ways and probably was even more similar a few billion years ago. Evidence suggests that there may have been a time when life as we know it could have flourished. The solid composition of Mars matches that of Earth with silicon and iron, though the overall density is lower, about three-quarters that of Earth. The planet lacks Earth-like tectonic activity. Data acquired by the *Mars Global Surveyor* spacecraft in 2005 suggest that activity may have occurred in the martian past, but planetary scientists are still debating the best interpretation.

Mars has a number of volcanoes and magma flows. Without tectonic activity (moving plates on the surface), these volcanoes are unlike the volcanoes that occur when plates collide. Mars cannot transfer heat by cycling magma as on Earth. Internal heat simply increases in some locations until it punches up through the surface. Hawaii shows similar processes on Earth, where magma broke through the middle of the Pacific plate. On Mars, this type of volcano was the only way to relieve all the excess heat in the interior. One giant eruption gushed out over the planet and formed the Tharsis bulge. Four huge mountains remain as a result of this activity, including Olympus Mons. The largest volcano in the Solar System, Olympus Mons exceeds sixteen miles in height. Its three companions, Arsia, Pavonis, and Ascraeus Mons, each exceed twelve miles. By comparison, Mount Everest on Earth only reaches up about five and a half miles.

The atmosphere of Mars includes mostly carbon dioxide (95 percent) with significant amounts of nitrogen (2.7 percent) and argon (1.6 percent). Although the high percentage of carbon dioxide might suggest greenhouse warming, the thinness of the atmosphere means that the carbon dioxide has little effect. Because of its small size, the planet cannot maintain a heavy atmosphere, and many of the volatiles like water have

　　　　　　　　　　　　　　　　　　　　　　　　　　LIFE IN SPACE

been lost to space over time. The remaining atmosphere is less than a hundredth as dense as that on Earth.

Mars tilts about 25°, which leads to Earth-like seasons. Carbon dioxide freezes into the polar caps during the winter and evaporates into the air in the summer. This causes a fluctuation in the density of the atmosphere of up to 25 percent over the year. With so little atmosphere, the planet experiences a wide range of temperatures, heating up in the day and rapidly cooling at night. Temperatures on Mars have been measured between −225°F and 95°F.

Scientists remain fascinated by the possibility of liquid water on Mars. Life as we know it requires water, so it should be a good marker for habitability. Current conditions on Mars suggest that liquid water could not occur on the surface. The low pressure and low temperature mean that ice would sublime—go straight from solid to gas—rather than melt. Water may be hiding in subsurface reservoirs, however, and stream forth for limited periods. In 2002, the *Mars Odyssey* orbiter detected large quantities of water-ice near the surface at the southern hemisphere. Channels on the surface suggest the presence of moving water in large quantities. In 2006, *Mars Global Surveyor* provided pictures of gullies on the surface that could only have been produced by liquid water flowing within the past seven years. Theories exist for how climate and erosion may reveal underground ice and cause sudden floods, but specific details are still a mystery. The water probably represents flash floods that quickly evaporate and may not provide an adequate medium for the development of life. If life arose in the distant past, though, these flash floods might be enough to maintain organisms adapted to such environments.

Astrobiologists continue to investigate whether the current situation—frozen water with occasional flash floods—was always the situation on Mars. Some surface channels on Mars could not have been produced by flash floods or geologic activity. The *Opportunity* rover made two interesting discoveries in several craters on Mars.[5] Rock layers on the sides of the craters look like the patterns inside sedimentary rocks on Earth. Sedimentary rocks form gradually at the bottom of bodies of water as particles settle down to the bottom. Sedimentary rocks on Mars would imply that a body of water was present for hundreds of years at some time in the past. In addition, *Opportunity* has found countless tiny spheres of hematite on the surface. These tiny spheres, or "blueberries" as they have come to be called, strongly resemble accretions found on Earth in highly acidic streambeds. Together with rippling patterns in the stone of the surface, the layers and blueberries make a compelling argument. Acidic, salty, flowing water must once have covered large regions of the plain where *Opportunity* landed. In the last chapter, we noted that the Sun has been very slowly cooling off over the past few billions of years. A warmer Sun in the past means a warmer Mars. Planetary scientists do not think the difference was enough to bring Mars up to Earth-like temperatures, but it may have been enough to support liquid water in the warmer regions of the planet.

Overall, Mars looks like a good candidate for past life. Water, carbon, and energy were all present in useful amounts. If life arises quickly wherever resources allow, then Mars probably had life once. Unfortunately we have no way to assess how common the origin of life might be. This is one of the questions we took with us to Mars in the first place.

Another important question to ask in the exploration of Mars has to do with whether past life would have left evidence. Astrobiologists disagree about evidence for life in the first billion years of Earth history. Mars history must be even more contentious. The most unambiguous evidence of life on Earth comes in the form of structural fossils, which probably did not start forming on Earth until after life had been around for at least a billion years. If life did arise on Mars, it may not have been around long enough to leave that kind of evidence.

Exploration of Mars

Human exploration of Mars stepped up significantly late in the last century and a complete list of Mars missions would not fit here.[6] The earliest explorations by the United States and the Soviet Union went by the names *Mariner* and *Mars,* respectively. In 1975, NASA landed spacecraft at Chryse Planitia and Utopia Planitia; the *Viking* landers returned important data about surface conditions. Three experiments specifically designed to detect the presence of life produced inconclusive results. It was not clear whether the reactions they detected required the presence of life and, in the absence of definitive proof, many scientists gave up on looking for life on the Red Planet.

Two important events occurred in 1996 that helped renew interest in Mars and hope for finding life there. One involved years of planning, whereas the other was quite sudden. In late 1996, NASA began a new wave of Mars exploration by launching *Mars Global Surveyor* and *Mars Pathfinder.* In 1992, NASA launched the *Mars Observer,* but lost contact with the spacecraft before it entered martian orbit. That meant that the new missions would return the first new data in twenty years. *Pathfinder* was also the first rover mission to explore another planet. Not only did it send back fascinating pictures of the surface and new experimental data, it acted as a test run for later rovers, allowing NASA engineers to design better and more durable models for future use. The rover module, *Sojourner,* returned data from the Ares Vallis region for nearly three months, twelve times the planned mission time.

The second event of 1996 involved evidence for life found in the Allen Hills meteorite ALH84001. Discovered in Antarctica in 1984, the meteorite has just the right composition of elements and isotopes to be from Mars. The chemical makeup suggests that the rock crystallized around 4.5 billion years ago with carbonate formations inside from roughly 3.9 billion years ago. The meteorite probably began as a piece of debris knocked off the planet's surface by another meteorite.

ALH84001 was originally mislabeled and researchers were unaware of its martian origin. Not until the 1990s did scientists take a serious look at it to find out more about Mars. In 1996, a team of scientists led by David McKay at Johnson Space Center announced that they had found something extraordinary. They saw evidence for life in the meteorite—martian life. The evidence eventually rested on fifteen separate arguments for biological influence on the materials. Notable among these were tiny packets of carbon resembling Earth fossils. The packets looked like fossilized bacteria found on Earth, though they were an order of magnitude smaller than known bacteria (closer to 30 nanometers across than 300). Other evidence came from chemical fossils known as polycyclic aromatic hydrocarbons (PAHs)[7] and strings of magnetite crystals resembling those found in some Earth bacteria.

More recent studies lead most astrobiologists to believe the evidence is insufficient. No one debates the martian origin of the meteorite. Life on Mars, on the other hand, represents a huge claim. Scientists want proof just as large and ALH84001 does not provide it. All fifteen lines of evidence presented by McKay's group have been individually questioned. In each case, abiological methods could produce the properties observed. Were ALH84001 a terran rock, I think that many microbiologists would be convinced that a new form of life had been found. Biological processes seem a more elegant explanation for what the McKay group found. Still, as long as a reasonable chance exists for life-free formation, that explanation must be favored. We will have to wait a little bit longer for proof of life on another planet.

After 1996 Mars exploration became a greater priority. Missions leave Earth approximately every two years as Mars' orbit brings it close to Earth. Missions failed in 1998 and 1999,[8] but 2001 saw the launch of the *Odyssey* orbiter, which returned high-resolution maps of the surface essential for later mission planning. In 2003, NASA sent two rovers to explore the martian surface. Larger versions of the *Sojourner* rover, *Spirit* and *Opportunity*, both exceeded expectations. They lasted eight times their expected lifetime and returned thousands of images. The first explored Gusev Crater while the second landed in Terra Meridiani. As I write these words, *Opportunity* is still moving and doing experiments on Mars, more than three years after arrival. In 2006 *Mars Reconnaissance Orbiter* reached Mars and began taking precise measurements of the surface, atmosphere, and magnetic field. Images from orbit have helped scientists direct the rover to the most interesting spots. NASA and ESA both have plans for future missions.

Deimos and Phobos

Mars has two tiny moons. Their names, Deimos and Phobos, come from the Greek words for dread and fear, attendants of the god of war.[9] Neither moon has sufficient mass to be spherical. Phobos is only around eleven kilometers across and Deimos about six. They have no significant effect on the environment or habitability of Mars. The Soviet Space Agency attempted to put a lander on the surface of Phobos, but was unable to do so. Information about the moons comes mostly from spacecraft orbiting Mars. Both moons have similar composition to objects in the asteroid belt and may be stray objects caught up in the planet's gravitational field.

The Asteroid Belt

Between the orbits of Mars and Jupiter lies a ring of small, rocky objects called the asteroid belt. During the formation of the Solar System, this region of space was never cleared out, and the objects that remain may be the same as the original planetesimals. Gravitational interactions with Jupiter and possibly Mars kept larger objects from forming. Larger objects caught up in the wake of Jupiter would reach high velocities and explode when hitting other large objects.

Most asteroids can be found orbiting the Sun at distances between 2.1 and 3.3 AU. Their composition reflects the history of planet formation: heavier elements closer to the Sun. Iron and silicon make up most of the inner asteroids, with increasing proportions of carbon and volatiles farther out. The largest known asteroid—and the first to be discovered—was named Ceres after the Roman goddess of agriculture. Ceres is a dwarf planet, spanning 975 by 909 km. That makes it roughly one-sixth the size of Earth by radius, or one-half the size of the Moon. Other asteroids vary in size from Ceres down to tiny particles.

It would be impossible to find, name, and track all the tiny pieces of rock traveling around the Sun. Most reside in the gap between Mars and Jupiter, but they can be found throughout the system. Three groups have attracted particular attention as a result of their regularity or possible interactions with Earth. Near-Earth asteroids (NEAs) represent the set of asteroids whose orbits intersect the orbit of Earth. Astronomers try to keep close track of large NEAs that have the potential to cause disasters by hitting us. A second group of asteroids travels within the orbit of Jupiter in two large clusters. One precedes the planet 60° ahead while another follows 60° behind. These asteroids are called Trojans and resemble the outer asteroid belt in composition. The final group, called Centaurs, move around the sun in orbits between Saturn and Uranus. Centaurs contain mostly volatiles like ice and closely resemble the comets in the Kuiper belt.

The small size of asteroids prevents them from having an atmosphere, or even enough mass to hold a puddle of liquid water. Life would probably be impossible. Asteroids, however, have played an important role in the history of life. Not only do they provide information about other planets, as in the case of the Mars meteorite ALH84001, but they have also changed the course of life on Earth. The Chicxulub meteorite, for example, crashed into the planet along the Yucatan Peninsula 65 million years ago. The 10 km piece of rock collided with Earth with an impact of 10^{14} tons of TNT. By comparison, hydrogen bombs produce on the order of 10^6 tons. The Chicxulub impact caused massive earthquakes and threw tremendous amounts of matter into the atmosphere. This event corresponds with the final days of the dinosaurs and the end of the Cretaceous period. The last of the dinosaurs and half of all known species became extinct during this period of disaster, cold, and darkness. One impact dramatically reorganized the biosphere.

Asteroids constitute an important part of the Solar System. They affect orbital dynamics, shape planets, and even affect life on Earth. Though few of them have recognizable names, astrobiologists consider them an integral part of the system in which life developed. The United States, Europe, and Japan have all sent spacecraft to collect data from asteroids. The U.S. *Near-Earth Asteroid Rendezvous (NEAR)* even landed a probe on the surface of the asteroid Eros in 2001.

Terrean life arose within a dynamic environment where the Sun, Earth, Moon, and meteorites all affect local conditions. Rocky planets and meteorites interact in the inner Solar System, giving Earth seemingly unique properties. Neither too hot, like Mercury and Venus, nor too cold, like Mars, Earth has just the right location to nurture life. Constant cycling of crust and atmosphere circulate elements, and the circular orbit and dense atmosphere provide stability.

Humans have investigated much of the local neighborhood, landing probes on Venus, Mars, and asteroids as well as gathering data remotely. These missions have provided key information about the origins of planets and life. As our exploration of the universe advances farther and farther into space, we will continue to learn more about the nature of the stars and planets and spread the influence of terrean life.

10

The Outer Solar System

At first glance, the outer Solar System might seem a poor place to look for life. The energy budget must be much smaller, given the additional distance from the Sun. The carbon composition of planets and moons must be low. Liquid water gets harder and harder to find. The giant planets themselves cannot support life as we know it. The incredible density felt at depth would be too high, and it can be difficult to imagine organisms living in environments of gaseous and liquid hydrogen. Minerals and heat necessary for life would be deep within the planet, but sunlight and more reasonable pressures would be nearer the surface. The planets themselves appear to be poor candidates.

Moons, on the other hand, might provide ideal places for the origin and development of life. They have more reasonable mass and composition. They even have interesting interactions with their parent planets that provide for energy. As we look at individual cases, we shall see how special situations might provide the water, electrons, and carbon we associate with life.

Jupiter

Jupiter is the largest planet in our Solar System, more than three times as massive as Saturn, and over three hundred times as massive as Earth. It travels around the Sun at an average distance of 5.2 AU. Mostly composed of hydrogen (90 percent) and helium (10 percent), the density of Jupiter is considerably lower than that of the rocky planets. Planetary scientists think that beneath the vast layers of clouds, a rocky core ten to fifteen times as massive as Earth anchors the planet. Measurements are impossible at that depth, however, meaning that scientists must base their models on the planet's mass (observable from gravitational interactions with other planets) and composition (determined by comparison with other planets and minor planets at various distances from the Sun). Above the solid core rests an ocean of liquid hydrogen. Hydrogen can only exist in this form under very high pressure—something like 400 times the pressure at sea level on Earth. The hydrogen exists as a collection of protons and electrons mixed together like a fluid, a situation found only in Jupiter, Saturn, and the Sun. The ocean probably also contains helium and trace amounts of other elements.

Different models have been developed for Jupiter's atmosphere between the ocean and space. Earlier theories have had to be readjusted in light of recent observations by the *Galileo* spacecraft and Earth-based research. Temperatures near the surface are around −243°F or 120 K. The highest layer of the atmosphere probably consists primarily of ammonia (NH_3) crystals. Going deeper into the planet, composition and temperature change as pressure and heat rise. A second layer seems to be composed primarily of ammonia and hydrogen sulfide (H_2S), and the third and lowest layer has more water. Recent observations suggest that far less water exists within the atmosphere than was previously assumed. Far from a calm layering of gases, the jovian atmosphere consists of constant storms far stronger than any hurricane on Earth. The difference in heat between the equator and the poles as well as differences in temperature and pressure between layers cause immense storms. The Great Red Spot is the best example; the color change delimits the edges of a giant storm hundreds of years old.

With incredible pressure and huge storms, Jupiter may not be the best place for life to exist, but it contributes to habitability in the Solar System in at least one important way. The gravitational effects of the giant planet have stabilized the regular orbits of other planets. All eight planets have nearly circular orbits around the Sun, preventing them from running into each other. Jupiter also helps habitability by collecting stray objects. Many asteroids that wander out of the asteroid belt get caught in the gravity of Jupiter, crashing into the planet, getting caught as moons, or trailing along with the Trojans. Jupiter's moons, discussed below, may also show great promise in the search for nonterrean life.

Jupiter has had a fascinating history with profound effects on the ordering of the Solar System. Four giant moons, known as the Galilean satellites, have been seen from Earth for hundreds of years. Galileo (1564–1642) first discovered the fact that they orbit Jupiter. He used this observation to challenge the commonly held belief that all objects in space revolve around Earth. A number of spacecraft have visited Jupiter and its many moons. In the 1970s, NASA sent *Pioneer 10* and *11* as well as *Voyager 1* and *2* to the giant planet. They returned pictures before passing on to the outer Solar System. More recently, the *Galileo* spacecraft, launched by NASA in 1989, settled into orbit in 1995 and collected data for eight years before finally falling into the atmosphere. The ESA mission *Ulysses* passed by Jupiter briefly during this period as well before moving on to investigate the Sun. More recent measurements were taken by *Cassini* (NASA, 1997) on its way to Saturn.

Jovian Moons

Jupiter has a collection of natural satellites, forming its own system. Unknown to most, the planet possesses rings like those around Saturn, though much less dramatic. The rings are small and difficult to see unless the planet and the Sun are properly aligned. As of 2006, the planet has sixty-three known moons, most of them named for lovers or daughters of Zeus.[1] The four largest moons—Io, Europa, Ganymede, and Callisto—are quite large and were observed at the beginning of the seventeenth century. The moons are near to the Moon in size, though they tend to be less massive because they contain so much water and other volatiles.

Io, the closest of Jupiter's four largest moons, resembles the Moon in both mass and size. Io resembles the rocky planets in composition, with an iron core and high amounts of silicon. This probably results from its proximity to Jupiter during formation. Proximity to Jupiter makes it a far more dynamic object than any other planet or moon. Tidal forces—caused by the gravitational effects of Jupiter and the other moons—deform Io as it passes through its orbit. The process heats the moon and constant volcanism results as the surface crust fails to keep in the sloshing magma underneath. The surface can change up to a hundred meters between high and low tide in some places. Other than Earth, Io is the only object in the Solar System known to have currently active volcanoes. Eruptions wipe out all evidence of meteor impacts; no craters can be seen on the surface.

Io seems to have too much energy to make a good habitat for life. In addition to volcanism, the moon gets energy from the intense magnetic field of Jupiter. As the moon passes through this field, tremendous amounts of energy are generated. The forces strip nearly a ton of mass from the surface every second, creating a cloud of charged ions around the moon. Life as we know it would quickly dissolve on the surface of Io.

The second nearest of the Galilean satellites is Europa. In many respects it resembles Io, but its composition is radically different. Europa possesses large quantities of water, either as ice or as a liquid. Massive tides warm Europa, though they are only about one-tenth as strong as the ones affecting Io. On this moon, the tides probably maintain an extensive ocean under the surface ice. Several lines of evidence point to liquid water. Planetary scientists were surprised to discover that very few large craters exist on the surface. Some process must be resurfacing the planet on a regular basis, leading many to believe that liquid water spills out. Second, the *Galileo* spacecraft detected a magnetic field induced by Jupiter. Some conducting material must be present around the inside of the planet. Liquid water with salts seems the most likely solution. Finally, the mass and size of Europa are consistent with an ice and water mixture.

Unfortunately, Europa exploration will prove difficult. Engineers are working on plans to reach the subsurface ocean, but currently there are no workable plans to place a drill on the surface that could tunnel through the many kilometers of ice. Looking at all the evidence, planetary scientists believe that Europa consists of a rocky core accounting for less than half of the depth of the moon. Over this core rest many kilometers of liquid water and ice. Different models give different depths of surface ice over the ocean. Further studies of composition, salinity, and surface features will lead to a better understanding, but a large saltwater ocean seems probable. Tidal forces from interactions with Jupiter and other moons would keep the water liquid despite the exterior temperature of 103 K. Craters would be eliminated whenever meteors pierced the surface ice or tidal forces formed giant cracks. Saltwater from underneath would spread out and freeze, making a new top layer in the area around the event.

The importance of liquid water to life causes astrobiologists to be very excited about Europa. Liquid water appears to be rare and here we have an ocean of it. Nonetheless, a number of other considerations are important. Carbon might not be difficult to find, as long as some form of internal circulation rotates heavier materials from the rocky core out into the water. Energy could be more difficult, though. Solar radiation hits the icy surface where life is unlikely to survive. Energy would have to be generated from some other source under the ice, possibly volcanic or magnetic processes. Some astrobiologists believe that Europa could have magma vents like the deep-sea vents at the mid-oceanic ridges on Earth. Such vents could provide both carbon and energy for an ecosystem.[2]

Europa also produces a thin envelope of gaseous oxygen, making it one of very few satellites to have an atmosphere.[3] The three icy moons of Jupiter all produce small amounts of oxygen from their surfaces. Solar radiation and small particle impacts scour the top layer of the ice, forming gaseous water. Chemical reactions divide the water into hydrogen and oxygen. The hydrogen, being extremely light, drifts off into space, and a layer of oxygen remains.

Ganymede and Callisto, the remaining large satellites, are slightly larger than Europa. They orbit Jupiter at a greater distance, meaning that they experience minimal tides and magnetic effects. Like Europa, they consist largely of water. Unlike Europa, they have thick icy surfaces covered with craters that suggest a more stagnant internal structure. On both moons, ice is mixed with silicon and other heavier elements. Ganymede has a number of mountains and other surface features, though their geologic source remains unclear. Callisto, on the other hand, has ancient surface features suggesting almost no change in the past 4 billion years aside from the formation of craters.

Saturn

Saturn resembles a smaller, colder version of Jupiter. Matter tends to be much farther apart in the outer Solar System. Not only was the initial (protoplanetary) disk thinner, but the planets are larger and clear out more of the local space. Saturn orbits the Sun at 9.54 AU, nearly twice as far out as Jupiter. The composition resembles that of Jupiter, with a rocky core surrounded by liquid hydrogen and a predominantly hydrogen and helium atmosphere. Extremely high winds move the upper atmosphere.

Several spacecraft have visited Saturn. NASA's *Pioneer 11* passed by Saturn in 1979 as did the two *Voyager* missions in 1980 and 1981. The three missions returned thousands of photos in addition to data on the magnetic fields and local environment before passing beyond the Solar System. In 2004, the *Cassini/Huygens* mission reached Saturn and began sending back data. A joint endeavor of NASA and ESA, the spacecraft has focused on Saturn science with a particular interest in the moon Titan and the rings. In 2005, the *Huygens* probe separated from *Cassini* and dropped into the atmosphere of Titan.

Saturnian Satellites

Saturn's most spectacular feature comes from the countless objects orbiting the planet. Within a certain range of every planet, gravitational forces shatter moons into tiny fragments.[4] At Saturn, the forces produce a spectacular array of tiny satellites forming giant rings around the planet. A number of gaps have formed within the rings as tiny moons clear out orbits. Most of the moons, however, can be found outside the rings. The NASA Jet Propulsion Laboratory (JPL) Web site lists fifty-six separate moons, though newly discovered objects regularly receive names as astronomers map their orbits. Thirty-five of the moons have proper names and half of those are named after Titans.[5] The most massive moon has been named Titan. The large number of satellites results in occasional rearrangement of smaller moons, as gravitational interactions pull them into new orbits.

Most of the saturnian satellites lack the size to be considered candidates for life; Titan, however, shows some promise. The largest moon in the Solar System, Titan has almost twice as much mass as the Moon. Observations of Titan's mass suggest a rocky core with iron and silicon as well as water-ice and solid methane. Atop the rocky core rests an atmosphere ten times as deep as the atmosphere on Earth. Curiously, that atmosphere may be very similar to the atmosphere on Earth before the cyanobacterial revolution (see Chapter 12) oxidized it. The major component is nitrogen gas (N_2, 95 percent) with methane (CH_4) as the next most common ingredient. Methane molecules in the atmosphere interact with each other and incoming radiation, forming a number of organic molecules. (Recall that organic only means that carbon-carbon bonds exist. No evidence exists of life on Titan.) Measurements place the surface temperature at $-290°F$ (94 K), making Titan colder than the coldest nights on the Moon or Europa.

Until very recently, the surface of Titan had been a mystery. The atmosphere near the surface is clear, but 120 miles up, it becomes opaque. Pictures were generated using radar imaging from the *Cassini* spacecraft and data from the *Huygens* probe as it settled to the surface. The surface clearly has texture, but few craters, indicating some form of resurfacing. Volcanism and tectonic activity have both been considered. Observations in 2006 lead planetary scientists to believe that lakes of liquid methane have settled into various low spots. Astrobiologists have become very excited that Titan may provide new insights into planetary chemistry, if not the origin of life.

How does Titan measure up for our three considerations, water, carbon, and energy? Water should be present in the rocky core, even though the surface temperature is too low for it to ever be found in liquid form. Life on Titan would need to utilize liquid methane as a medium. Carbon exists in abundance, both in the atmosphere and in the ground. It seems easy to imagine sufficient quantities accumulating in methane lakes. Energy presents a trickier question. The atmosphere allows a great deal of light in. A person on the surface would experience brightness equivalent to 350 times the light available on a bright moonlit night. That light could be turned into usable energy, but organisms would need to absorb large quantities in order to combat the extreme cold. It remains unclear whether or not such life could exist, but it presents some interesting possibilities. Now that we have a better idea of conditions on Titan, experimental chemists will be able to simulate the environment, and discover how molecules interact under those conditions.

Uranus

Passing beyond Saturn, we find ourselves in the newer part of the Solar System. Humans have known about the first six planets for thousands of years, but the outermost planets were only discovered recently. In 1781, a British astronomer named William Herschel (1738–1822) noted a very small blue disk through his telescope. The fact that he saw a disk—rather than a twinkling point of light—suggested that it was not a star. Herschel's observations led to the discovery of the seventh planet, Uranus, and two of its moons.

Uranus has far less mass than Saturn and Jupiter. Hydrogen (83 percent) and helium (15 percent) make up almost all the matter in the planet, making it about one-quarter as dense as Earth. Methane (2 percent) and trace amounts of water and ammonia are also present. About 80 percent of the mass forms a liquid core—like a giant, three-dimensional ocean. Deep within that ocean, heavier materials probably exist, but in relatively small quantities. A deep atmosphere covers the surface of the ocean. The top layer of the atmosphere contains mostly methane gas, which gives the planet its pale blue color.

Uranus is far too cold to support life. It orbits the Sun at 19.19 AU, twice as far out as Saturn, and the surface temperature rests around 59 K (−350°F). The planet does have a number of curious properties, however, that would make it an interesting case study for planet formation. Unlike all the other planets, Uranus rotates about an axis lying in the plane of the planets. If we looked at the Solar System on its edge, Uranus would appear to be lying on its side, spinning up and down (like a bicycle tire) rather than sideways (like a top). The magnetic field has formed around an axis 58.6° off of its rotational axis, meaning it lines up with neither solar north nor planetary north. To make matters worse, the magnetic axis does not run through the center of the planet, but about one-third of its radius away from the center. Something interesting must be going on inside.

Only *Voyager 2,* launched by NASA in 1975, has passed close to Uranus. The spacecraft flew by the planet in 1986, collecting numerous images and measuring the magnetic field. Planetary scientists have great enthusiasm for studying the odd behavior of the seventh planet, but recognize substantial engineering challenges for working so far from Earth and the Sun.

Uranian Satellites

Uranus' system includes twenty-seven known moons and a spectacular set of rings, surpassed only by Saturn. Because of the planet's strange tilt, the rings point out of the plane of the planet's orbit, making it possible, on a rare occasion, to see them all at once from Earth. (Saturn's rings, lying in the orbital plane, will always be partially obscured by the planet.) In addition to the rings, there are a number of moons, mostly named after Shakespearean characters.[6] The largest of the moons, Titania, approaches

Earth's Moon in size, but it is far less dense. Astronomers think that the most massive of the uranian satellites consist of a half-and-half mixture of rock and water-ice. None of them appear to be good candidates for finding life.

Neptune

Neptune comes next as we travel toward the outer edges of the Solar System. It is slightly smaller than Uranus, but a little more massive due to a higher concentration of helium (80 percent H_2, 19 percent He, 1.5 percent CH_4). Neptune orbits the Sun at a distance of 30 AU, making it extremely cold by our standards (48 K). It resembles Uranus in composition and structure, but lacks the strange rotation and magnetic field patterns. It lacks sufficient heat and the heavy elements necessary for life. The planet was discovered in 1846 by Johann Gottfried Galle (1812–1910), a German astronomer looking to explain oddities in Uranus' orbit. Astronomers had predicted the presence of an eighth planet based on the effects it had on the seven known planets.[7] Even today, almost all observations of Neptune have been made from Earth-based telescopes, although *Voyager 2* passed by in 1989.

Neptunian Satellites

Neptune has faint rings as well as thirteen named moons. The largest satellite, Triton, was discovered by Galle in the nineteenth century and a second, Nereid, was found a century later in 1949. The *Voyager* spacecraft located six additional moons and the rings. The moons were named after the children and attendants of Poseidon, the Greek god of the sea.[8]

Triton is nearly as big as the Moon and one of the densest objects in the outer solar system. Only rough estimates have been made for Triton's composition, based on long-range measurements made by *Voyager*. Water-ice with some other frozen volatiles probably make up most of the moon's mass. Ice and volatile geysers punch through the icy outer crust, regularly recovering the surface. The ice and organic dust make the surface highly reflective, and this results in temperatures below 35 K (around −400°F). Triton has two claims to fame. First, it travels around Neptune in the direction opposite the planet's spin; the only large moon to do so. Second, Triton's density suggests that it has far more rock in its interior than any other moon around Saturn, Uranus, or Neptune. The two facts lead astronomers to believe it was generated elsewhere in the Solar System, migrated, and was captured by Neptune.

The Kuiper Belt

Out past Neptune, we find trillions of miles of debris. Small clusters of frozen volatiles orbit the Sun at tremendous distances. Although a few larger objects have coalesced, most of the objects resemble tiny dust particles separated by so much space that they never interact to form planetesimals. Closer to the orbit of Neptune—from perhaps 35 to 50 AU—the particles still stick to the planetary plane. Formed out of the outermost fringes of the protoplanetary disk, these objects constitute the Kuiper belt.[9]

When we think of comets, one image comes to mind: those objects that rarely pass by Earth, leaving bright trails behind them. Because of their heavy concentration of volatiles, comets slowly dissolve when traveling through the inner Solar System as the solar wind excites their surface. Many of the volatiles stay nearby, forming a halo or corona around the solid core of the comet. Other particles get blown outward, forming a long tail that points away from the Sun. These comets have irregular orbits that cause them to swing close in by the Sun and then far out into space. They can form spectacular shows visible from Earth. The vast majority of comets, however, exist in the outer reaches of the Solar System, where solar radiation is less intense. Like asteroids in the asteroid belt, they slowly drift around the central star. Comets formed in the outer Solar System; they have almost no heavy elements, but consist of frozen volatiles like hydrogen, helium, water, and methane.

LIFE IN SPACE

No spacecraft have measured objects out beyond the orbit of Neptune, although nine missions have focused their attention on comets closer in.[10] In 2006, NASA launched the *New Horizons* mission and that spacecraft is currently passing Jupiter. It is expected to arrive at Pluto in 2015.

Pluto caused much controversy recently as astronomers debated whether or not it should be called a planet. Scientists had long recognized that it orbited the Sun with a number of other Kuiper belt objects (KBOs), but Pluto held pride of place as the largest. Noting discrepancies in the calculation of Neptune's orbit, astronomers began looking for a ninth planet in 1905. Twenty-five years later, Clyde Tombaugh (1906–1997) found the tiny object. Subsequent studies suggest that Pluto, named after the Roman god of the underworld, does not have a noticeable impact on the orbit of Neptune, but it is relatively large—about two-thirds as wide as the Moon—and has a satellite, Charon. Controversy over what it should be called began as soon as it was discovered, although it has been considered a planet for the last seventy-five years. The recent discovery of several massive KBOs, including at least one larger than Pluto, has made many less certain. In 2006, the International Astronomical Union declared that a planet must be an object that has cleared out local space. Pluto has not, meaning that it is now considered a "dwarf planet." It shares Kuiper belt space with at least three other large bodies.

Pluto's orbit is far more irregular than any other planet's orbit. As it travels around the Sun, it passes as close as 30 AU (within Neptune's orbit) and as far away as 48 AU. The planet's composition appears similar to that of Neptune's moon Triton, with a rocky core covered by much ice. All observations have been made from Earth, leaving planetary scientists with minimal data.[11] Spectrographic analyses (looking at the wavelength of light reflected by an object) suggest that Pluto has a surface made up of methane, nitrogen, and carbon monoxide ice.

Pluto has its own satellite, Charon.[12] Charon has about one-eighth the mass of Pluto; astronomers, therefore, tend to say that the two objects orbit one another. Both objects dance around each other as they pass around the Sun. Charon looks slightly less dense than Pluto and appears to have water ice on its surface. During the summer, as Pluto passes closer to the Sun, the surface melts. A thin atmosphere forms that can reach out as far as Charon, allowing particles to travel back and forth. During the winter, the atmosphere freezes over again.

In 2003, a team at Palomar Observatory near San Diego California discovered a new and very large KBO.[13] Originally named 2003 UB313, the object now bears the official name Eris after the Greek goddess of discord. Eris' orbit stretches from 38 to 97 AU, taking 560 years to make a complete circuit of the Sun. Best estimates size the object at 2,400 km across, making it slightly bigger than Pluto. Mass and composition estimates have proven more difficult. One thing we do know is that Eris reflects a great deal more light than Pluto, suggesting a smooth, white surface. This may be a result of finding the planet in the middle of winter, when the atmosphere has frozen solid. Summer measurements will have to wait another 280 years. Like Pluto, Eris has its own satellite. The satellite takes the name Dysnomia, daughter of Eris and spirit of lawlessness. We still know very little about this moon.

Over 1,000 KBOs have been identified in the past fifteen years. Two approach Pluto and Eris in size. Awaiting further knowledge, they still hold the provisional names 2003 EL61 and 2005 FY9. Both are about three-quarters the size of Pluto and travel in similar orbits (between 35 and 52 AU). Astronomers are particularly fascinated by 2003 EL61 because of its irregularity. The object resembles a football and spins around its short axis every four hours. Two small moons have already been observed as well. Other named KBOs include Quaoar and Orcus.[14] Future research on KBOs promises to reshape our understanding of celestial mechanics.

The Oort Cloud

Farther out than even the Kuiper belt, a corona of dust known as the Oort cloud surrounds the Sun in all directions. These particles were captured by

LIFE IN SPACE

the Sun's gravity or remain from the original formation of the star—objects that almost, but not quite, escaped. No direct observations have been made, but our experience of comets works well with the theory that these Oort cloud objects orbit the sun in random trajectories. Interactions with extra-solar objects occasionally sling a comet into the Solar System. This theory would explain why so many comets do not line up with the plane of the planets.

Astronomers suspect that the Oort cloud stretches roughly from 2,000 to 100,000 AU. The only such object observed has been named Sedna. At present we know almost nothing about it. Observations since 2004 show Sedna to be near the height of summer, passing around 75 AU from the Sun. At midwinter (12,000 years from now) astronomers expect the object to reach nearly 1,000 AU. The temperature on Sedna should be around 33 K with a severe cooling off expected as it passes away from the Sun. Best guesses on size say that Sedna should be half as big as Pluto. It may also have a satellite.

Farther from the Sun, the Solar System looks less hospitable to life. With less energy coming in and fewer heavy elements available, the giant planets would not support life as we know it. On the other hand, water is far more abundant, if mostly in the form of ice.

Several moons in the outer Solar System provoke questions for astrobiologists. Europa and Titan in particular show interesting chemical processes and may even be good environments for life. Gravitational and magnetic fields provide energy and dynamism to Europa, and a thick atmosphere covers Titan. At the edges of the Solar System, as we come to the end of the Sun's influence, we begin to see different types of objects that have made us redefine how we think of planets and moons. In the next chapter, we reach out even farther to consider planets orbiting other stars.

11

Extrasolar Planets

Earth is not as unusual as we once thought. Within the past three decades, astronomers have discovered more than 300 planets beyond our Solar System. The search has only just begun, but every newfound planet tells us something interesting about the way planets work. Every new planet shows us one more example of the basic principles of life. Above all, every new planet presents a new opportunity for life.

The majority of these planets will probably turn out to be giant balls of ice, dust, and gas—much like the objects in our own outer Solar System. Right now technology limits our detection and we mostly see very large objects orbiting close to their parent stars. Astronomers have labeled these "hot Jupiters" because they probably resemble larger versions of that giant, orbiting within a few astronomical units of the central star. Technology improves, though. Every year we find more objects and objects that fit more closely with our ideas of habitability.

Astronomers have reported on 306 extrasolar planets.[1] The vast majority were inferred based on radial velocity (RV) measurements, which look at subtle, periodic shifts in the quality of starlight. Astronomers have also observed planets by looking at star brightness (transit photometry), location in the sky (astrometry), and rotation (pulsar timing). A few planets have even been imaged directly.

Radial Velocity

In 1989, David Latham and colleagues reported a fascinating discovery.[2] Careful measurement of light from the Sun-like star HD114762 showed periodic shifts in the quality of its light. As stars move away from us, their movement stretches out the light we receive, just as the frequency of a police siren decreases as the police car speeds away. Latham's research group looked at the periodic shifts and asked a question: What if those shifts represent the movement of the star as another star swings around it? What if we see the shifts because the star moves slowly back and forth?

Patience proved the team right. The light shifts corresponded to a pattern. HD114762 swings back and forth in an eighty-four-day cycle consistent with the presence of a very close and very large companion. Latham's group originally suggested a brown dwarf star. The expected mass seemed too low, but that might just reflect a high margin of error or the angle of observation (see below). The object seemed to be 11–13 Jupiter masses with an orbit similar to that of Mercury. Very large, very hot. Subsequent discoveries led researchers to question whether or not it could be a planet; it remains on the edge of definition. Whether we consider it a failed star or a superplanet, the object around HD114762 started our current list of extrasolar planets. The original object goes by the title HD114762 b.

The method was not used again for detecting planets until 1995. Michel Mayor and Didier Queloz of the Geneva Observatory found a companion object for the star 51 Pegasi.[3] The companion object, 51 Peg b, could be only half as massive as Jupiter and orbits the star every 4.2 days. It presents a much more compelling case for the existence of an extrasolar planet and began the current rush to find new planets.

How do RV measurements work? Imagine a heavy ball at the end of a short rope. By grabbing the end of the rope and swinging the ball around your head quickly, you can keep the ball in the air, but it pulls you back and forth as it swings. RV measurements look at the stagger of the central star (you) as it swings a satellite (the ball) around in space. If we get lucky, we will see a star at just the right angle. If our line of view lies within the orbital plane, then the staggering will occur directly away from us as the planet passes on the far side of the parent star. The light coming from the star toward Earth will become redder, because the star is moving away and the light is stretched out. The staggering comes right back toward us as the planet swings back around between Earth and the parent star. The light becomes bluer as the star moves toward us and contracts the wavelengths. By taking careful measurements on a regular basis, astronomers can fit a curve to the various wavelengths and calculate the frequency and magnitude of the stagger. How often does it occur? How far does it go?

RV measurements present certain difficulties for astronomers. The greatest problem has to do with inclination. If we truly see directly into the plane of the planet's orbit, then our measurements will reflect the true mass and orbital distance. On the other hand, if we are above or below the plane, then the stagger to and from Earth only represents a fraction of the full stagger. We would then underestimate the planet's mass.[4] Even worse, we might miss the planet altogether if the plane of orbit is perpendicular to our line of sight. Then we would see no stagger at all—at least not by measuring radial velocity.

Another issue for RV measurements arises from the need to see a recurring pattern. Latham's team found their planet (or star) because it caused a regular stagger every eighty-four days. Thus, if we wanted to catch an Earth-like orbit, we would need to detect a regular pattern over a period of years. More distant planets like Neptune and Uranus may require decades of careful observation because they take so long to make a single pass around their parent star. Multiple planets can make the situation even more complicated as astronomers need to account for stagger caused by many interacting spheres all flying about the same central star. Each one affects the others' position. The more planets, the more drunken the star will appear, and the harder it will be to find an underlying pattern to the stagger.

LIFE IN SPACE

Finally, we should bear in mind that only the most massive planets will create an observable stagger in their parent star. Earth, at less than 1/300th the mass of Jupiter, produces a far more subtle signal than most planets observed so far, though our techniques improve every year.

Even bearing the difficulties in mind, RV measurements have opened up a large chunk of the galaxy for planetary science. Few of these planets have been observed directly, but astronomers have inferred their existence on the basis of light. Between 1989 and mid-2008, 289 planets in 248 systems have been predicted in the region near Earth. Keeping in mind that those only reflect large, hot planets close to their stars, orbiting in the right plane and having caught the attention of astronomers, it's an amazing number. How many more planets have yet to be found?

Astrometry

A similar method of extrasolar planet detection should work if the observer looks in perpendicular to the orbital plane. The staggering of stars as their planets pull from various directions moves them in a tiny circle in the night sky. To give you some idea of how small these movements are relative to Earth, consider the night sky. A star would be expected to move on the order of one microsecond of arc through the sky as a result of planetary forces. That corresponds to 1/3,600,000,000th of a degree for observers on Earth. Astronomers attempted to find stars this way as early as the 1950s, but were never able to make their measurements precise enough. Even today, telescopes lack the resolution to make astrometry a reliable source of planet detection. The Hubble telescope did manage to record the stagger of Gliese 876 in 2002, confirming the presence of a planet already detected by RV measurements.

The true promise of astrometry comes from its ability to detect planets with large orbits. They cause the most significant stagger. A long-term observation plan, measured in decades, should be able to detect planets invisible to other methods. This would fill in gaps in our knowledge and lead to a better understanding of smaller, more distant objects. In the meantime, other methods will need to suffice.

Pulsar Timing

The first time astronomers knew they were discovering a planet occurred in 1991. Aleksander Wolszczan and Dale Frail used precise measurements of radiation coming from the pulsar PSR1257+12 to predict the presence of three orbiting objects.[5] The behavior of pulsars, and the radiation they emit, is so regular that it is (relatively) easy to detect the slightest irregularities in their rotation.

A very small percentage of stars, after exploding as supernovae, spend the end of their lives as pulsars. An extremely dense cinder remains after the death of a star. The fuel for nuclear fusion has been spent, but the incredible compaction of mass from the star's life continues to show some activity. In the case of a pulsar, that activity comes in the form of gigantic jets of matter and radiation emitted from the magnetic poles as the star spins. Just like Earth, the pulsar's magnetic pole and axis of rotation are not the same. This means that the jets move with the rotation of the planet and cause a lighthouse effect. When viewed from another planet, the light from the jets blinks on and off as the star rotates. Pulsars were first described in 1967 because of a blinking radio signal.[6] Subsequently, astronomers have found the beacons radiating light over a huge variety of wavelengths.

Wolszczan and Frail used slight irregularities in the pattern of light from PSR1257+12 to predict the existence of three orbiting planets. The amazing consistency of pulsar flashes allows astronomers to detect objects as small as one-tenth of an Earth mass. Astronomers refer to the new objects as PSR1257+12 b, c, and d. The two larger objects have a minimum mass nearly three times that of Earth and rotate about the star in 98 and 67 days, respectively. The smaller object was not as well characterized. In 1994, Stephen Thorsett and colleagues discovered a fourth pulsar planet orbiting the star PSR B1620-26.[7] This planet is much bigger, more than twice the mass of Jupiter, and orbits every one hundred years. Only last year, another giant planet was discovered, V391 Peg b.[8] So far, only these five planets have been discovered using pulsar timing techniques.

Despite the wonderful accuracy of pulsar timing and its ability to detect very small planets, it has limited utility to astrobiologists. We know too little about planetary systems to say for sure, but it seems unlikely that a planet would survive the supernova stage of a star. Even if it did, its orbit would have to change radically during the process. Does that mean that these five planets were somehow acquired by their parent stars after the explosion? Possibly. The existence of pulsar planets creates a whole new suite of questions for planetary scientists. A second problem arises when we ask whether we could expect life around a pulsar. The irregularity of radiation and the coolness of the star suggest not. Whatever life we might find would look nothing like life as we know it.

Transit Photometry

The third method to produce results in the quest for extrasolar planets was transit photometry. In 2002, a team of scientists at Caltech and Harvard were the first to discover a planet using this method. Maciej Konacki, Guillermo Torres, Saurabh Jha, and Dimitar D. Sasselov reported a periodic dimming of the light coming from the star OGLE-TR-10[9] and attributed the dimming to a small object passing between it and Earth. The object has about 90 percent of the mass of Jupiter and passes in front of the star every 1.2 days.

Transit photometry has incredible benefits over the other methods. When astronomers know exactly what they are looking for, they can make measurements of the size and composition of a planet based on how light from the parent star passes around it. Just as a green filter will change the color of light passing through, the atmosphere of a transiting planet subtly changes the characteristics of light for its parent star. Close estimates of radius and planetary density can also be made based on the amount of light the planet blocks as it passes.

These benefits have matching challenges. Planets, being so much smaller than stars, block out only a tiny fraction of the light. The brightness of a star may diminish by only one percent in a transit, making it essential to be looking at the right place at the right time. Detection can also be complicated by poor observing conditions, because we are trying to detect such small changes in brightness. The most significant drawback comes from the inclination of the planet. Astronomers will only observe a transit if the planet comes directly between the star and Earth. We have to be within ten degrees of the orbital plane to see any signal at all. Transit photometry yields the best results when coupled with radial velocity measurements, and most planets have been verified by additional evidence.

To date, fifty-two planets orbiting fifty-two different stars have been detected using transit photometry. They range in mass from one-third to twelve Jupiter masses and have orbital periods of one to six days. Transit photometry is the most informative and second most successful form of extrasolar planet detection.

Direct Imaging and Microlensing

Two new methods for extrasolar planet detection arrived on the scene in 2004: direct imaging and microlensing. Both methods look promising, but rely on newly developed and still developing technologies. They provide unique information about extrasolar planets but require rare combinations of events to be useful.

Direct imaging does exactly what it says. Astronomers captured images of four planets by filtering out the light of their parent stars and catching the reflected light of orbiting planets. Gael Chauvin and associates at the European Southern Observatory took a picture of the brown dwarf 2M1207.[10] High-resolution infrared imaging shows a small irregularity in the star's circumference consistent with a planet with the mass of five Jupiters orbiting 46 AU from the star. Two planets were captured on film in 2005, one in 2006, and one more in 2007. These planets were even larger and farther away from their parent star and may represent brown dwarfs.

Microlensing takes advantage of rare celestial alignments. When a star passes between Earth and another star, a process called microlensing takes place. The gravity well of the nearer star bends the light streaming by in a way that magnifies the light of the distant star. Large planets distort the regularity of the lens and disrupt the image. Astronomers can use the disruption and the expected magnification to calculate the characteristics of the planet. Bohdan Paczyński proposed the idea in 1991, but undisputable results were not reported until 2004. The Optical Gravitational Lensing Experiment (OGLE) has so far found six planetary microlensing events. Microlensing promises to provide information about planets of size and orbit similar to Earth, but the process relies on a unique alignment of stars. The lensing star must pass directly into the line of sight between Earth and the distant star. The planet must also be in just the right position orbiting the lensing star to affect the final image.

Both microlensing and direct imaging promise to expand our understanding of planetary systems. Although neither technique works commonly at present, future technology should improve our chances. Direct imaging will become more useful as imaging resolution improves. Microlensing will continue to become more applicable as we chart the paths of more stars.

What Kind of Planets?

The next question we might ask is this: what kind of planets are out there? At the very least, what kind of planets have we found? Since most were discovered by RV, we can assume that they will be biased in the same way that RV measurements are biased—toward very large, very hot planets. Indeed this is the case. With the exception of PSR1257+12 b, which orbits a pulsar, the smallest extrasolar planet found to date is at least three times as massive as Earth. Of the 306, all but twenty are more massive than Neptune and thus expected to be gas giants. Even in the habitable zone, pressure at the surface would make Earth-like life improbable.

Still, there is hope. Life as we know it may not occur in the depths of a gas giant, but it could arise on one of that planet's moons. Even more exciting is the possibility that we have only seen the outer parts of some very interesting systems. The Solar System includes four rocky inner planets and four gas giants. Perhaps the giants we have observed around other stars have companion planets too small to notice. Three scientists in the United Kingdom have modeled the movements of known stellar systems and found that at least half of them would allow for the presence of a rocky planet within the habitable zone.[11]

In spring of 2007, an even more promising possibility arose. Gliese 581, a red dwarf about twenty light years from Earth, hosts a fascinating system with at least three planets. Gliese 581 c, the smallest of these, orbits just inside the habitable zone with a predicted temperature between 0 and 40 degrees Celsius. A team of astronomers led by Stephanie Udry at the University of Geneva has been monitoring the system. No one knows whether the planet has water or any of the structural components necessary for life, but it presents intriguing questions.[12]

What else can be said about the known extrasolar planets? RV detection favors large planets close to the central star. Figure 11.1 shows the properties of known extrasolar planets relative to planets in our Solar System. More than a third of the planets have less mass than Jupiter, and the number decreases steadily as mass increases. This suggests that the vast majority of orbiting objects will be small, just as in our Solar System. Only technology limits our observation of the smaller objects. Just more than half of the planets have orbital periods of less than a year, and almost all have orbits less than ten years. This reflects the fact that shorter orbits are easier to observe in a limited time using current methods. Roughly half of the planets orbit their parent star closer than Earth orbits the Sun and all but a few orbit within 5 AU. The preference for speedier orbits surely biases detection toward closer planets.

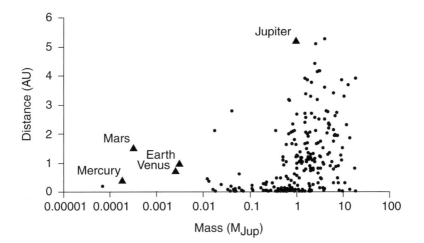

11.1 Extrasolar planets. Mass and distance of all reported extrasolar planets as of June, 2007. Jupiter and the four rocky planets in the Solar System appear for reference. Masses are given relative to Jupiter; in other words, 0.1 means one-tenth the mass of Jupiter. Distances reflect the semimajor axis of reported planets. Six planets orbiting their star at more than 6 AU have been omitted, as have seven planets whose orbital properties are unclear.

Another important consideration—one less closely tied to observational bias—involves eccentricity. Some planets form an almost perfect circle as they pass around a star. Others form long ellipses, passing close to the star and then sailing out into the distance before returning. Most of the planets in the Solar System have eccentricities near 0. Other than Mercury ($e = 0.21$), they all have eccentricities below 0.1, meaning that they stay close to the same distance from the Sun throughout their orbits. This circularity keeps their temperatures fairly constant and decreases the likelihood that they will run into each other and destabilize the system. Astrobiologists expect this to be a common property for inhabited systems because it allows for long-term consistency.

Roughly one-quarter of the known extrasolar planets have eccentricity below 0.1. (Unfortunately, Gliese 581 c is not among them; $e = 0.13$). This suggests that the Solar System may be rare in its stability and consistency. Planets with higher eccentricities would have more severe seasons—perhaps passing in and out of the habitable zone every year. They would also be more likely to have irregular gravitational interactions with other planets, which could change their orbits dramatically in a short period of time.

When we look at planets as orbiting platforms for life, we begin to see the tremendous possibilities for life in space. Once it seemed improbable to find a planet of just the right mass and composition at just the right temperature and distance from just the right star—the Goldilocks planet. The abundance of extrasolar planets found in the last thirty years changes that. Astrobiologists have become confident that planets like those in the Solar System occur throughout the galaxy and throughout the universe.

Astronomers remain uncertain of just how many stars our galaxy contains. Estimates tend to be in the hundreds of billions. If even one in a million of those have planetary systems, and if even one in a thousand of those systems have rocky planets in the habitable zone, we could still expect hundreds of candidates for life as we know it.

12

Life and Time

We turn now from generic to specific, from universal principles to local application. What has been a galactic survey becomes a planetary study. Up until now, I have covered issues having to do with life and the possibility of life elsewhere. Even when we looked at our own Solar System, it merely set the stage. Now we turn to life here on Earth and start to meet the players. The next seven chapters look at issues specific to terrean life and how it works.

Astrobiology encompasses a variety of disciplines. We have seen that already in the interactions of chemistry, physics, astronomy, and geology. It should be even more evident in the chapters to come as we tie biology into what we already know. For this reason it will be necessary to revisit a number of topics, including entropy and the definition of life. We are now prepared to visit them at the intersection of astronomy and biology.

Entropy Returns

I said a little about life and entropy in Chapters 4 and 6, but now I will make specific connections between entropy, life, and metabolism. Life takes small, relatively uniform units and builds them into larger and more diverse assemblages. Information gets incorporated into larger and larger aggregations of matter. At the same time, the second law of thermodynamics is never broken. Entropy—the measure of disorder—increases. How can this be?

Imagine a volcano covered by pools of water. Two forces are at work. Gravity draws the water downward while the force of the waves casts some of it up, and the heat of the mountain makes it evaporate and rise. The foaming surf pounds against the base, casting spray against the side of the mountain. Most of the spray falls back into the sea, but a few drops lodge in the nooks and crannies. Irregularities in the stone catch the water and shallow pools form on the slope. Heat from the magma begins to boil off the puddles and each one spits up steam and sizzling spheres. Most of the water rolls down the mountain, but a few drops land higher up, forming new pools higher on the slope. Gravity always operates, but some of the water still moves up.

Life resembles this watery volcano, with the Sun acting like the sea. Vast quantities of energy strike Earth every second. Photons come in like the tide, pounding the planet with more energy than it could ever keep. Most of the energy gets reflected back into space, but some of it gets trapped in the biosphere. Tiny organisms catch photons and trap them in chemical bonds. These molecules act like cracks in the mountain, holding energy in place. The bonds that form have less energy than the photons. What is more, the bonded molecules face an energy barrier. As much energy as they store, even more energy would be required to break them apart. The matter cannot simply fall back into a less complex form. Increased complexity resembles the slope of the volcano. The energy of the Sun (like the tide and the heat of the volcano) forces matter up the slope and into places where it cannot fall down again. Astrobiologists are fascinated by each of the steps because each one shows how life ratchets itself up the mountain of complexity.

We find the process fascinating precisely because disorder increases. The vast quantities of energy produced by the Sun, the tremendous energy lost at every level of the food chain, energy expended in movement and reproduction, even energy expended in thought; all of these represent the tremendous loss of energy and complexity involved in the process. The tiny amount of light captured by a leaf, the fraction of energy gained by a deer eating the leaf, the tiny segment of biological individuality compacted into a gene or a genome; these represent only the smallest edge of all the energy streaming into space as a star burns through its hydrogen.

For me, the most fascinating aspect of biology lies in seeing energy trapped elegantly in the biosphere. How do self-replicating systems capture even this small fraction of stellar power? My own research involves photosynthesis, the trapping of light energy in carbon-carbon bonds. Chapter 16 explores the various methods for using energy to put carbon molecules together (and rearrange them), a process called metabolism. Even this process exists within a network of other reactions scaling from small to large. Biochemistry is the first level at which we can appreciate life as life, a level from which we can finally see how the chemistry of stars and planets interacts with that of organisms, communities, and biospheres.

Filling the Niches

The story does not end with metabolism. All of this energy trapped within the biosphere needs to be maintained. Amazingly, it appears to seek out more energy and new methods of storage. Entropy applies at higher levels as well, so that every organism exists by balancing the needs of preservation and reproduction against the cost of fuel. Ecologists refer to the role of an organism within its environment as its niche. The word captures both how it relates to the nonliving world in terms of space and available energy and how it relates to other organisms.

Individual species survive for one of two reasons. They utilize a unique resource or they outcompete rivals in using a common resource. The giraffe's long neck allows it to reach leaves too high for most other animals to reach. Likewise, cyanobacteria became incredibly successful over 2 billion years ago because they filled a niche nothing else could fill—they were able to live on water, carbon dioxide, and light. Other organisms required energy from other sources, such as reduced sulfur compounds, that were easier to use. Even today, most organisms with which you are familiar cannot get energy from sunlight. Plants depend upon cyanobacteria within their cells to generate energy from light. Animals and fungi digest carbon compounds, mostly produced by cyanobacteria and plants.

The lion, on the other hand, is an example of an organism that outcompeted other animals to fill its niche. The lion gets its fierce reputation from its ability to fight off competitors when hunting prey. Cats have well-developed muscles, teeth, and claws, all excellent for catching and defending meals. Humans also show an amazing ability to outcompete other species. We use intelligence, tools, and a highly versatile anatomy to protect our resources from other species. Consider our ability to keep insects away from our crops. The struggle for control of resources continues, but we survive because we compete successfully.

Each organism or group of organisms fills a particular niche in the biosphere. Their life and death mark the movement of resources within the system. In a rare case, an organism will die and all of its complexity is lost. In most cases, however, one organism's demise becomes a source of water, carbon, and electrons for a host of other individuals. Ecologists study what niches are available, what organisms fill those niches, and how all the niches fit together.

Presumably, there was a time when all life on Earth filled a single niche. The first organisms must have used the same resources in the same location. As life progressed, different organisms began to spread out and use different resources. Some stayed in the same place and competed for water, energy, and carbon in the traditional way. Others found new ways to survive. The biosphere must have spread as more energy got trapped within the biological system. New resources meant new opportunities to trap energy within the system—new chances to store a copy of your DNA in a different location. Variation ensured that each new generation explored a wider array of niches.

When we look at the history of terran life, we can see that organisms start small and get bigger; organisms start without nerves and intelligence only arises later. Does this mean that evolution moves in a direction? Surprisingly, no. It means that life spreads out to fill all the available niches. The first life was simple, small, and unintelligent. Life started at the base of the mountain of complexity. As it spread out, it began to fill niches higher up the slope, but it also expanded around the base. Like any mountain, the mountain of complexity has more room at the bottom than at the top. Organisms become more complex to utilize new resources, but they also evolve new ways to compete for available resources. Starting with one cell, an organism cannot get smaller, but it can get bigger. Thus we see the arrival of larger organisms and more complex interactions between organisms through time.

As Stephen J. Gould said, it was and is and will long be the "age of bacteria."[1] Whether you look at energy budget, location, or effect on the planet, no organism has a greater impact than bacteria. They are as highly evolved as primates, though they have evolved to fill different niches. Whereas dinosaurs and humans use size and intelligence to have greater control over their environment, bacteria have achieved such variety that they can live on almost every human surface, from our doorknobs to our digestive tracts. Humans remain small in number relative to the countless billions upon billions of bacteria that still compete for resources everywhere on Earth. Large size and high intelligence both represent extremely small niches within the biosphere, filled late because they were difficult to fill and because the rewards were small relative to the energy budget of the biosphere.[2]

How to Measure History

One last introduction will be necessary before we dive into Earth history in depth—an introduction to time. Time might seem like a familiar concept, but once we begin to deal with billions of years it can be difficult to grasp. How does one measure time on those scales and how does one relate life to it? Paleontology exists at the interface between geology and biology. The word paleontology comes from the Greek *paleo,* ancient, and *ontos,* being, and means the study of ancient things. Traditionally, paleontologists explored the past by looking at fossils, the structural remains of ancient life. Recent discoveries have allowed paleontologists to explore biomarkers, also known as chemical fossils, as well. Some metabolic processes produce molecules that cannot, to our knowledge, be produced abiologically. Complex carbon ring structures like steroids provide evidence for biological activity billions of years ago. Other biomarkers relate to the abundance of gases (oxygen, for example) and isotopes (such as carbon thirteen and carbon fourteen).

This section looks at some of the important moments in Earth history that allow paleontologists to map the evolution of terrean life. Figure 12.1 shows major landmarks geologists use to divide Earth history into manageable bits.[3] The history of Earth can be divided into four eons or ages of life, the Hadean, the Archean, the Proterozoic, and the Phanerozoic. The Proterozoic and Phanerozoic Eons have been further divided into periods with names like Cambrian and Cretaceous. All of these names reflect the types of fossils found from that time period.

The Birth of Earth

Planetary scientists now believe Earth to be roughly 4.5 billion years old. A number of lines of evidence lead to this approximation, but the most convincing method is called uranium-lead dating. In some elements, the number of protons and neutrons making up the nucleus is so large and poorly balanced that it wants to kick out some of the subatomic particles. In the case of uranium (U), the atoms end up losing ten protons and a number of neutrons to form lead (Pb). Chemists call this process radioactive decay because the larger atoms (U) decay to form smaller atoms (Pb) and produce radiation. No one can predict when an individual atom will decay, but large numbers of atoms will behave in a regular way. Radioac-

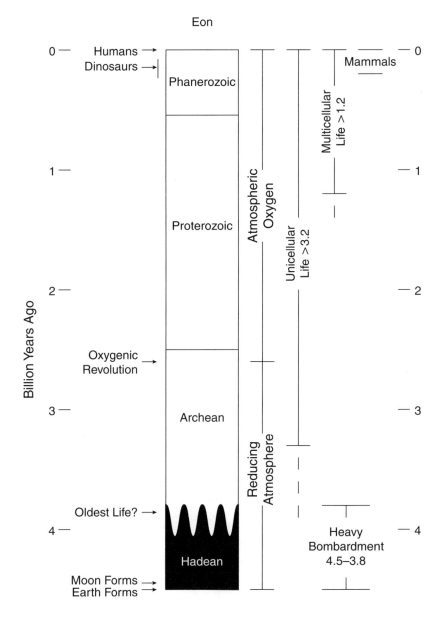

12.1 Earth history.

tive elements decay with a half-life, $t_{1/2}$, so that in every $t_{1/2}$ years half of the atoms decay. One isotope of uranium, ^{235}U, decays to lead, ^{207}Pb, with a half-life of 704 million years.[4] That means that half of the ^{235}U will turn into ^{207}Pb after 704 million years. In twice that time, 1.4 billion years, three-quarters of the ^{235}U will turn into ^{207}Pb; that is, one-half in the first 704 billion years and one-half of the remainder in the next. Knowing this, paleontologists can measure the amount of ^{235}U in certain minerals and see how long it has been sitting there by how much remains relative to the ^{207}Pb.

To be sure that they have measured correctly, geologists use uranium found in a mineral called zircon. Zircon typically contains the elements zirconium, silicon, and oxygen. Conveniently, zircon formation occasionally incorporates uranium, but regularly rejects lead. This means that lead found within the zircon must represent decayed uranium. Uranium dating also comes with a built-in correction system. One isotope of uranium, ^{235}U, decays to an isotope of lead, ^{207}Pb, with a half-life of 704 million years, but another isotope, ^{238}U, decays to ^{206}Pb with a half-life of 4.46 billion years. A calculation using ^{235}U can be double-checked using ^{238}U, all within a single zircon crystal. Uranium-lead dating of meteorites and moon rocks suggests that the Solar System, at least in the vicinity of Earth, reached its current configuration approximately 4.54 billion years ago.

The Hadean: Hell on Earth

The first eon of Earth history has been named the Hadean, after Hades, the Greek underworld. The name represents a conflation of Hades, a shadowy realm, with much later conceptions of Hell as a fiery furnace. During the Hadean, the surface of Earth may have included oceans of molten rock. The planet was still cooling down from its original formation from accreting planetesimals.

The Hadean must be the most mysterious part of Earth history, partly because of its great distance in time. Originally describing the time before formation of the oldest rocks geologists could find, the Hadean stretches forward from the origin of the planet for 500 to 700 million years. At some point plate tectonics started, rearranging Earth's crust, and erosion and other factors destroyed everything earlier. More recent discoveries, however, make this definition less useful. Currently, the oldest known rocks come from the Acasta Gneiss in Northwest Canada (nearly 4 billion years old) and the Isua formation in western Greenland (around 3.4 billion years old). Some debate exists about whether or not the timing of the Hadean should be pushed back so that no Hadean rocks can be found. No matter how you look at it, though, very few Hadean rocks exist. Zircon crystals, which survived intense geologic activity, persist from the earlier periods (up to 4.3 billion years old), but, with few exceptions, the oldest formations are no older than 3.8 billion years.

The Hadean Eon would have been extremely hostile to life as we know it. The surface of the planet was pulverized by multiple meteorite impacts as Earth cleared its orbit of leftover debris. Meteorites streamed down from space in what is known as the "heavy bombardment" period. One immense impact with a Mars-sized object around 4.44 billion years ago resulted in a rearrangement of Earth's rock and the ejection of matter to form the Moon. Some planetary scientists have suggested that regular impacts would have had enough power to completely vaporize any water on the surface of the planet as recently as 3.9 billion years ago. Other scientists find this infeasible. The notion is connected to an idea called "impact frustration." Repeated meteorite impacts would have such an effect on surface conditions that life could not form until the environment settled down. However you feel about impact frustration, general opinion holds that life was not present in the Hadean.

The Archean: The Origins of Life

When did the earliest life arise? This has been the subject of heated debate among astrobiologists. The question hangs upon how you define life and what you consider sufficient evidence. The best contender, in my opinion, rests on isotope evidence from Akilia Island in western Greenland.

Isotopes are not only useful for radiological dating, they also show evidence of a process called "fractionation." Various chemical processes favor one isotope over another. For instance, one isotope might be more likely to react or might freeze just a little bit more easily. Photosynthesis preferentially includes the lighter ^{12}C over ^{13}C when assembling carbon molecules. Although the first isotope is naturally far more common than the second, photosynthesis will further refine the available carbon, decreasing the ratio $^{13}C/^{12}C$ in tissue and biological remains. Oil and limestone, both remnants of ancient organisms, contain lower than average $^{13}C/^{12}C$ ratios.

In 1996, Stephen Mojzsis and colleagues reported that the isotope ratios of sedimentary rocks on Akilia Island were consistent with biological fractionation.[5] If the 3.85 billion-year-old formation truly represents evidence of life, it would mean that life predates the geologic record. Does this mean that life arose as soon as the surface of the planet cooled down? Possibly. Reconstructing events over 4 billion years ago can be difficult, but it seems startling that life could have formed so quickly. A few scientists have questioned whether the rock formation came about in the method proposed by Mojzsis. Extremely high temperatures could also account for isotope fractionation, though that explanation has problems as well. Currently, the most plausible answer is that simple organisms, perhaps resembling modern photosynthetic bacteria, lived at the beginning of the Archean Eon.

Astrobiologists know little about this earliest period of life. Several lines of evidence suggest that the early Earth was quite warm. Geological evidence supports temperatures from 55 to 85°C (131–185°F) early in the eon. Nitrogen was the major component of the atmosphere. Oxygen would have been scarce and the atmosphere was thick with carbon dioxide, methane, and water. As the Archean passed, the climate became more temperate. Hydrogen was lost to space, meteor impacts slowed down dramatically, and water condensed to cover large portions of the surface.

The Archean Eon stretches from the first rocks until the rise of atmospheric oxygen. No evidence exists for land-based organisms until the very end of the eon. All life was unicellular, anaerobic (existing without oxygen), and aquatic.

The Proterozoic: Simple Life

Around 2.5 billion years ago the terrean atmosphere changed. Oxygen levels began to rise dramatically. Paleontologists believe that the new oxygen came from a biological innovation—oxygenic photosynthesis. Five groups of bacteria utilize sunlight to make energy (phototrophy), but only cyanobacteria are capable of separating hydrogen from water in the process. Cyanobacterial, or oxygenic, photosynthesis builds carbon-carbon bonds using carbon dioxide and strips water into its component atoms. Bacteria pass hydrogen through membranes to generate energy and oxygen is released. The rise of oxygenic photosynthesis meant that life was no longer limited to areas where resources were easily available. Cyanobacteria could live anywhere that sunlight, water, and carbon dioxide were available. In other words, they could live anywhere on the surface of the planet. This revolution must have rapidly spread across the face of Earth, transforming local biospheres into one gigantic ecosystem and radically expanding the reach of life. The expansion produced huge volumes of oxygen gas as a byproduct and pumped them into the air.

The oxygen had two profound effects on terrean life. First, it acted as a terrible poison. We associate oxygen with life, but oxygen can be highly reactive. Organisms that evolved prior to the revolution probably had no defenses against this new toxin. Second, the readily available reactive oxygen allowed whole new metabolisms to develop. More carbon was incorporated into the biosphere and organisms could utilize respiration to convert sugars to energy. Some biologists have even suggested that multicellular life could not have arisen prior to the oxygen revolution.

Exact dating can be difficult. Chemical evidence for cyanobacteria-specific metabolisms occurs as early as 2.8 billion years ago. Structural evidence predates even that, reaching back to 3.2 or 3.5 billion years. Oxygenation of the atmosphere and oceans occurred between 2.45 and 2.22

billion years ago. Oxygenic photosynthesis may have arisen early and taken hundreds of millions of years to have a discernable impact. Alternatively, most of the cyanobacterial metabolism may have been in place early, but genuine oxygenesis was not happening until later. In either case, the world changed dramatically.

The Proterozoic marks the period of time between the rise of oxygen and the rise of multicellular organisms. Life was increasingly aerobic, predominantly aquatic, and mostly unicellular.

The Phanerozoic: Visible Life

The current eon, the Phanerozoic, was created to describe the time in which visible life occurred. Rocks from this period contain recognizable fossils of multicellular plants and animals. All commonly known extinct species, from trilobites to dinosaurs, lived during the Phanerozoic. For this reason, the first three eons often get lumped together under the heading "Precambrian."[6] Within the Phanerozoic, time has been divided into three eras—Paleozoic, Mesozoic, and Cenozoic.

A lengthy ice age occurred at the end of the Proterozoic Eon and, when the ice receded, new species abounded—including the precursors of modern plants and animals. The Phanerozoic officially begins with the Cambrian Period 542 million years ago, but careful study of fossils, both structural and chemical, shows that the first stages of multicellularity extend back well into the Proterozoic. Eukaryotes big enough to be seen with the naked eye began appearing around 1.8 billion years ago. Red algae from 1.2 billion years ago show evidence of photosynthesis, multicellularity, and sexual reproduction; making them remarkably similar to plants.[7] For perspective, consider the age of dinosaurs, which lasted from roughly 230 to 65.5 million years ago.

The species that surround us today only appear gradually throughout the Phanerozoic. Some debate exists around the origins of flowering plants, which include all flowers and grasses and most trees. They arose sometime in the Mesozoic Era and did not become truly common until after 65 million years ago. Today they dominate the environment. Mammals (animals with spines, warm blood, and hair; including humans, cats, bears, and so on) arose approximately 195 million years ago. Like the flowering plants, they did not dominate until after the Cenozoic Era.

The four eons make it easier to see the sweep of biological evolution and put it all into context. The Hadean Eon stretches from the formation of the planet until things cool down enough to leave a record. The Archean Eon, showing evidence of life from its earliest years, includes the period from the end of the heavy meteorite bombardment to the rise in oxygen levels. The Proterozoic Eon covers the period from oxygenation of the atmosphere to the rise of abundant multicellular life. Multicellularity no longer marks a clear division at the end of the Proterozoic, but the ice age and the Cambrian explosion do.[8] The Phanerozoic Eon, from the Cambrian onward, marks the period of abundant visible life, with multicellular plants and animals taking on increasing importance across the globe. The history of life is one of change and expansion, from the first unicellular organisms to the incredible diversity we see today. The environment has shaped that change as life filled the various niches, but terran organisms have also had a profound impact on the planet they inhabit.

13

Making Cells from Scratch

Life as we know it resembles a skyscraper made out of building blocks. Smaller components come together to form more complex entities. One of the most fascinating things about life is that this appears to be true at every level. Like a fractal pattern, the same pattern appears at different scales: A small number of interchangeable parts can be assembled in a nearly infinite number of ways to make something larger. Part of defining life will be figuring out how life arises in scale as well as how life arises in history.

Where do we begin, and what are the smallest building blocks? Current science only goes so far down the spectrum.[1] At present the smallest known level contains elementary matter and elements of force. These are quantized; in other words they come in indivisible packets. Quantum physicists study these tiny packets—electrons, muons, taus, and quarks, which make up matter; gravitons (hypothetical), gluons, W and Z particles, and photons, which make up force. Of course, at such a tiny level, matter and energy resemble one another more closely than they do in the macroscopic world. Packets can behave like waves or particles. This makes them extremely difficult to measure. With few exceptions, elementary particles are too small to have a meaningful size using units we would find familiar. Quantum physicists tend to rely on particle mass when considering size. Still, these nine types of packets can be assembled to compose most known physical phenomena.[2]

Subatomic Particles

The next level involves subatomic particles—electrons, protons, and neutrons. Subatomic particles are measured in femtometers (10^{-15} m). Note that electrons occur here and at the elementary level. Partially this results from historical accident. Electrons were discovered far earlier than quarks. They are often dealt with in atomic structure alongside protons and neutrons despite having 2,000 times less mass. The take-home message is that building blocks do not always interact at their appropriate level. Sometimes blocks from level 1 and 2 will come together to form level 3 structures.

Quantum physicists call second-level building blocks composite particles. Protons and neutrons are both composed of an assembly of three quarks; the difference results from quark flavor. Two "up" quarks and a "down" quark form a proton. Two "down" quarks and an "up" quark form a neutron.[3]

Atoms (Isotopes and Ions)

Protons, neutrons, and electrons come together to form atoms. The third level of organization is the atomic level. Ancient Greek and Indian philosophers proposed a smallest, indivisible element of matter at least 2,500 years ago. The philosophical position called atomism makes the claim that some level of matter is the most fundamental and no lower level exists. The word atom derives from the Greek *atomos,* unable (*a-*) to be cut (*-tomos*). When chemists were beginning to work out the periodic table in the nineteenth century, the word "atom" became attached to this level of matter. The name stuck, despite the fact that smaller levels were subsequently found. Physicists measure atoms in fractions of a nanometer ($0.1-0.5 \times 10^{-9}$ meters).

Every atom has two parts, a nucleus and electrons. The nucleus forms a dense core made up of roughly equal numbers of protons and neutrons, held together by the strong force. Around this nucleus swarm a number of electrons. The negative charge of the electrons holds them close to the positively charged protons in the nucleus. All atoms follow this pattern, from the simplest (hydrogen = 1 proton and 1 electron; helium = 2 protons, 2 neutrons, and 2 electrons) to the most complex (for example, uranium = 92 protons, 141–146 neutrons, and 92 electrons).

Of course the number of neutrons and electrons can vary. Proton number determines the name of the atom. One proton means hydrogen. 92 protons means uranium. Various numbers of neutrons result in isotopes. There are, for example, four common isotopes of carbon, carbon-11 through carbon-14. Each isotope has 6 protons. Neutron number ranges from 5 (carbon-11 = 6 protons plus 5 neutrons) to 8 (carbon-14 = 6 protons plus 8 neutrons). As we saw in Chapter 13, isotope chemistry can be extremely important in dating the origins of Earth and terran life.

Various numbers of electrons result in ions. Ideally, the charge of every atom is balanced. Protons have positive charge. Electrons have negative charge. If they occur in the same numbers they balance each other out. This does not always occur, however. If an electron gets lost, the atom will acquire a positive charge (e.g., H^+). If an electron gets added, the atom will acquire a negative charge (e.g., Cl^-). Tracing the movement of electrons will be essential to understanding biochemistry and metabolism. Electrons are the most basic currency for energy in life, so we need to be able to know exactly where the electron is and what the charge is on important atoms. Anytime an atom has a charge, it can be referred to as an ion. Just to confuse matters, biochemists will often refer to a hydrogen ion (H^+) as a proton, because once you remove the electron, nothing else remains.

If we look back at the chapters on cosmology, stars, and planets, a few things should now become clear. Why is there so much hydrogen and helium in the universe? Because they are so easy to assemble from small parts.

Why are stars and gas giants made up of hydrogen and helium? Because they are abundant in the universe. Why are supernovae important to astrobiologists? They form immense pressure chambers wherein all the heavier elements can form. Carbon and oxygen both form only in the death throes of stars, meaning that all three basic conditions for life as we know it—carbon, water, electrons—come from stars.

Molecules

Chemists call groups of atoms joined together with chemical bonds—or unattached atoms—molecules. Atoms form the building blocks for molecules. Molecules form the fundamental unit of chemistry as they exchange electrons or, rarely, as they exchange protons and neutrons (nuclear chemistry). This can be a confusing level of organization, because molecules come in a wide range of shapes and sizes. A simple hydrogen ion floating in the cell (10^{-10} meters and one atom) constitutes a molecule, but so does a DNA chain (up to 10^{-7} meters and tens of millions of atoms). Biochemists make a further distinction of levels for clarity. Simple units, called monomers, constitute the most basic molecular building blocks. A small number of monomers in four categories form all known life: nucleotides, amino acids, simple sugars, and fats.

Organisms string monomers together in long chains called polymers. There are three fundamental polymers of life: nucleotides form nucleic acids (DNA and ribonucleic acid [RNA]), amino acids form polypeptides (which make up proteins), and sugars make up carbohydrates. Fats aggregate and can be incorporated into the three types of polymer, but form no simple polymers of their own. We should finally have reached a level of familiarity. Individual organic polymers may not be visible to the naked eye, but long-chain polymers such as proteins and carbohydrates make up our daily diet.

Inorganic polymers also exist. PVC should be a common example— polyvinyl chloride. Chemists can make chains of PVC of incredible length and then attach the strands to make a durable solid plastic. In some sense, a PVC pipe is a single molecule, with connections between all the different atoms.

Between a Molecule and a Cell

The next two levels of organization seem slightly arbitrary to me. Atoms and molecules form the fundamental building blocks of chemistry, but cells form the fundamental building blocks of biology. In between them lie several levels of cooperation and integration that form the realm of biochemistry. The first of these I have called "functional units," both because the unit size is functional and because the units have particular functions in the cell.

Proteins form the prototypical functional unit. Proteins do something. Most proteins consist of a single polymer, a polypeptide, though many proteins use several. A polypeptide is nothing more nor less than a long string of amino acids. Polypeptides can be generated in a laboratory with an almost infinite variety from around thirty common amino acids.[4] A protein consists of a single polypeptide, or a combination of polypeptides that act together within a biological system. Enzymes, proteins that catalyze specific chemical reactions, are a great example. You may have heard of lactase, the enzyme that allows people to digest lactose. The enzyme catalyzes a reaction separating lactose into simple sugars. Signal peptides are another example. Organisms produce both oxytocin and insulin and use them to transport messages around the body. A protein is a polypeptide with a purpose.

Chromosomes also constitute a functional unit. A long strand of DNA containing the genes for an organism can be called a chromosome. Once again, random strings of DNA can be assembled in a laboratory, but the word "chromosome" implies a particular function. In humans and many other organisms, chromosomes entail much more than a single string of nucleotides. For us, a chromosome involves huge numbers of packing proteins as well. The packing proteins allow great lengths of DNA to be curled up into a tight package and then selectively unfurled. Chromosomes are DNA for information storage.

As you might be able to tell from my language, we have now crossed over fully into the realm of biology. Abiological chemicals rarely warrant the notion of function or purpose. Insofar as they accomplish things, they do so solely as a result of energetic concerns. Thermodynamic laws drive reactions. Once in the realm of life, however, we begin to see things moving toward an end—specifically the maintenance and replication of organisms, keeping energy and order trapped within a system. In truth, biochemical reactions follow all the same laws that abiologic ones do. Biochemistry achieves local results at the expense of energy in the larger system, just as does solar chemistry. Neither shows evidence of intention, per se. Nonetheless, once we reach the level of functional units, it becomes easier to think about things as if they had a purpose.

Function in Biology

Just to be perfectly clear, I want to say a few things about purpose. As humans, we favor life—not a controversial statement, to be sure. Likewise, we have a very fond view of birth. Babies are good. The purpose we attribute to biochemistry from organisms down to functional units is the extension of life and the production of offspring. Over the last few centuries, biologists and biochemists fell into the habit of talking about purpose in a very specific way. They began to talk about the purpose of all proteins and chromosomes. They said that the primary function of functional units was to keep the host organism alive long enough to produce offspring. Even evolutionary biologists—perhaps especially evolutionary biologists—have a tendency to talk about purpose in this fashion.

Unfortunately, the common understanding of purpose can be very misleading. Many functional units operate to their own benefit—at the expense of the organism in which they are found. Chromosomes may exist with no other purpose than self-propagation at the expense of their host (e.g., plasmids). Some proteins like prions may actively harm their host. Purpose has so much infiltrated the way biologists communicate that it would be nearly impossible to remove purposeful language from a discussion of biology. It is important to remember, however, that "function" will always be a category applied by the observer. It is not necessarily an objective property of the system.

In summary, functional units consist of molecules assembled in a way that does work. This work might be information storage (as with DNA), catalysis of reactions (as with enzymes), information transfer (as with signal peptides), or any number of other functions. Functional units represent only our understanding of a dynamic set of interactions. Work accomplished may not always serve the ends we assume.

Organelles

Organelles form another somewhat arbitrary unit, but one very useful to understanding biochemical processes. Organelles are organlike subunits of cells. Cells with nuclei (eukaryotes), have a number of organelles, each of which localizes a set of biochemical processes. The nucleus, a membrane-enclosed region for storage and access of DNA, was one of the first named organelles. Others include the Golgi apparatus for chemical processing, the mitochondrion for energy production, and the chloroplast for photosynthesis. Mitochondria and chloroplasts represent another great example of mixing between levels. Microbiologists now know that these two organelles are descended from free-living bacteria. They may have started as parasites or they may have been eaten but not digested. However they came to be there, they represent a quasi-organism living within a cell.[5] Once again, we see building blocks at one level (organism/cell) being incorporated with elements at another level (organelles) to form a new living arrangement.

Cells

Cells have long been called the fundamental unit of life. Proteins and fats make up a flexible bubble that holds biochemical processes in place. Biologists refer to the bubble as the cellular membrane and the biochemically

LIFE IN SPACE

active interior as cytoplasm. For hundreds of years, biologists believed that cells make up all living things. Recently this position has begun to weaken. We now see selection and evolution occurring in odd ways, but familiar life always happens in cells. For the next few chapters, let us assume that all life is cellular.

Microbiologists and cell biologists measure cells on the micrometer scale (μm, millionth of a meter). The smallest known cell comes in the form of a bacterium, *Mycoplasma genitalium,* which can be as small as 0.2 μm in diameter. Currently, biologists cannot even imagine a smaller organism. Cell membranes, minimal amounts of DNA, and the necessary structures for the maintenance of life require at least that much width.[6] More common cells range from 1 to 5 μm. The largest cells come in the form of eggs. Eggs contain nutrition for the earliest stages of life, but begin with a single set of genetic information at their core. The largest ostrich eggs can reach 20 centimeters (cm) across.

Since the 1940s, cell biologists have divided cells into one of two types, eukaryotic or prokaryotic. The word "eukaryotic" comes from the Greek *eu* meaning "true" and *karyon* meaning "nut." Eukaryotic cells have a clearly defined darker area, an additional membrane-bound region that encloses most of the nucleic acids—a nut at their core. Prokaryotic, from *pro* and *karyon,* means prior to the nut; prokaryotic cells were believed to predate eukaryotic cells.

Bacteria have only one prokaryotic cell per organism. For this reason, biologists call them prokaryotes.[7] All prokaryotes have a flexible cell membrane and most of them have a stiff outer shell called a cell wall. The cell wall maintains the shape of the prokaryote and protects it from harm. Prokaryotes also have a nucleoid. Like the nucleus, the nucleoid represents a centralization of nucleic acids, but no membrane holds the nucleoid together.

Animals and plants consist of large collections of communicating cells, all of which are eukaryotic. Fungi and a wide array of one-celled organisms (e.g., amoebae) also possess eukaryotic cells. For this reason biologists call them eukaryotes. Most eukaryotic cells lack a cell wall, though plants produce them and use them to build skyscraperlike structures that dwarf all other organisms (e.g., the redwoods). Eukaryotes are characterized by a large number of internal structures—organelles. Most notable is the nucleus. The previous section mentions a few others. Paleontologists have long been fascinated by the question of how prokaryotes and eukaryotes are related and why multicellularity (many cells in one organism) appears limited to eukaryotes.

Cell Theory

Cells were first noted by Robert Hooke in the mid-seventeenth century. Hooke observed that cork was made up of tiny hollows. They reminded him of the rooms in which monks live, so he called them cells. Later, he saw similar structures in other plants. In the following years, biologists found that all multicellular organisms contained this same pattern of repeated units. They decided that cells formed the fundamental unit of life; all living things are composed of cells and nothing smaller can truly be called alive. With the advent of genetics, biologists expanded the definition further. All cells contain a complete copy of an organism's genetic complement. Cell theory was born.

Cell theory holds two major premises. The first states that cells make up the basic unit of life. Every organism consists of one or more cells. The second states that cells come from cells and only from cells. In Chapter 2, I talked about "spontaneous generation," the idea that life could come from nonlife. After Pasteur, biologists rejected this possibility. Cell theory gives the alternative more definition. Cells form the irreducible unit of life. One cell may divide to form two or more cells, but a cell may not arise by any other method. Mind you, this still does not explain how the first cell came to be, but it does get you from one cell to the tremendous diversity of cells, organisms, and communities we see today. So "cell" means more than a structural unit. It also has a component of information in it. Recent experiments have shown—at least in theory—that every cell of a multicellular organism contains enough information to construct a clone.

Some biologists still argue, based on cell theory, that anything below the level of a cell is not alive. If a line must be drawn between nonlife and life, they draw it here. This seems to me to be a useful, reasonable, and observation-based assertion. Based on that definition, viruses and a host of other anomalies discussed later clearly do not constitute life. They have no cells. They are not alive.

Personally, I prefer to put the definition of life a couple steps lower on the hierarchical scale. As I mentioned above, I think that proteins and genes have many properties we associate with life. They can evolve and function in ways that promote their own well-being while harming the cell. Our ideas of function clearly demonstrate some idea of life. Beyond that, I prefer the answer that allows for further explanation. If life strictly requires a cell, then I fear we will never be able to know more about the origin of life. The definition seems to prohibit meaningful questions about the nature of an entity in the process of becoming a cell.

A definitive answer may not be possible until we have a clearer picture of biochemistry, evolution, and the variety of life. The question, however, remains incredibly important for astrobiologists. As we struggle to explore the origin, extent, and definition of life, we must continually challenge our assumptions about what life is at the most fundamental level. The hierarchical scale allows us to identify our own assumptions, be clear about what we mean, and try to identify life within a larger framework.

14

Building Biospheres

What is an organism? The *Oxford English Dictionary* defines organism as "an individual animal, plant, or single-celled life form." Even if we disregard the matter of fungi and other nonplant, nonanimal, multicellular life, we must admit that the dictionary has thrown us back on two other definitions. We established earlier that life can be a bit tricky. Cells form discrete packets of space and information, but we know that we want something larger than a cell—an individual. We want a category that includes humans as individuals, not to mention dogs, cats, rutabagas, and any number of other entities we encounter on a daily basis.[1]

Individuals should span the range of creatures from the smallest bacterium to the largest redwood. Individual animals often have personalities and motivations. Mammals like chimpanzees, otters, and dolphins show evidence of reasoning. Even worms and beetles can be identified as individual organisms fairly easily. Amoebae and bacteria might be more complicated to identify, but when we think of infections, it is easy to imagine a single one-celled organism causing all sorts of problems. The smallest one-celled organism would be the smallest cell, being around 0.2 μm across. The largest living tree on record is called General Sherman, a coastal redwood (*Sequoiadendron giganteum*) almost 84 m tall and 31 m across at the base. Several claims can be made for the largest organism of all. Multiple tree trunks can sprout from the same root structure, or genetically identical colonies of fungus can spread for a mile underground. No matter how you measure it, organisms span a huge range of sizes.

What is an individual life form? In the case of prokaryotes and other one-celled creatures, we seem to be on solid ground. One cell, one organism. Problems really arise only with multicellular organisms like humans and trees. The most common definition involves genetic identity. Many multicellular organisms go through sexual reproduction, intentionally mixing up their genes every generation to produce more variety. For this reason, you represent a collection of genes, with only half coming from each of your parents. You are unique. This uniqueness can be understood by looking at your genes. The idea has become so popular that most readers will be familiar with genetic fingerprinting. By sampling enough genes from a cell, you can tell which person it came from. Each person has a unique complement that either matches or does not match a given sample. This genetic identity will form our working definition of individuality. An individual life form is the set of all cells sharing identical or almost indistinguishable genetic components.

The use of genetic identity can be extremely useful when looking at organisms. It allows us to group together families of organisms—such as you and your parents—and to distinguish between different organisms—such as a mother and the fetus growing inside of her. The field of phylogenetics involves tracing the ancestry of individuals by comparing their genetic elements. We can do paternity tests, reconstruct the transmission of diseases, identify criminals, and track the migration of animals, all using the genetic identity definition. In spite of what I am about to say, I want you to know that genetic identity makes a lovely and useful definition for an organism.

Alas, the definition only works about 99 percent of the time. Here is why. Genetic identity misses in a few very important cases. A colleague used to joke that voting should be run by genetic identity; identical twins should get only one vote. Identical twins happen when a single fertilized egg splits in two within the mother. Two genetically identical people develop from one genetic blueprint. Are they one individual or two? Clearly two. Their personalities, experience, and character are quite distinct. So genetic identity fails the twin test. We know that more than one individual can share a single set of genes.

The definition fails for humans in the other direction as well. A genetic condition known as chimerism results when two fertilized eggs fuse together within the mother. Two completely different genetic components may be expressed in a mottled pattern across the body. Skin color, eye color, even sex can alternate between different parts of the body. A chimera can be made up of little patches of different material. One person, one individual, two genetic identities. So genetic identity fails the individuality test as well. Studying only humanity, we can already see that genetic identity fails some of the time.

The question becomes even murkier when we study bacteria. Bacteria will often grow in large colonies of genetically identical individuals. Because bacteria do not have sex and do not reshuffle their genes between generations, all members of the colony have the same genetic makeup.[2] Should the colonies be considered individuals? They may even stay connected to one another and exchange materials. This is the case for some types of cyanobacteria (for example, *Nostocales*). Some species even reproduce as a group. What stops these groups of cells from being considered individuals?

Within biology, the organism is often referred to as the "unit of selection." This is another way of saying that evolution always selects for (or against) organisms. This definition immediately rules out the possibility of competition between groups and leaves us at a loss to deal with the selfish functional units described in Chapter 13.

With a few caveats, still we return to the common-sense definition of an organism as a genetically unique individual. The word can be quite helpful in understanding biology, and I do not wish to discourage people from using it. At the same time, we need to remember the limitations and biases it introduces to our thinking.

Communities

Communities can be even more problematic than individuals. They can be communities of common descent, such as species and families; communities of common interest, such as societies and symbioses; or communities of common location, such as ecosystems. In each case, a single entity has been constructed of individual organisms, and something interesting can be learned from looking at life through the lens of community. A community might be as small as two organisms living together and exchanging resources or as large as a rain forest. I bring up communities not just because they are the next step on our hierarchy of scale, but because several important issues in astrobiology revolve around how we use these ideas.

Communities of Common Descent

Communities of common descent should be familiar to you. For the past three centuries, biology has been organized around a set of nested categories. In the eighteenth century, a Swedish biologist named Carl Linnaeus began classifying organisms in a new way. Linneaus introduced binomial nomenclature, in which organisms are identified by two names, genus and species. Genus indicates a broader category (for example, *Felis*, cat), whereas species indicates more precisely the type (for example, *Felis catus*, the house cat).

Over the next century this system expanded and gained precision. Species came to refer to a group of individuals capable of interbreeding.[3] It was thought of as the basic unit of taxonomy (classification and naming of organisms). Biologists believed that it represented an unambiguous and essential element of individuals. Higher-level taxonomy arose in order to group species into larger and larger groups. The Linnaean system can tell you many interesting things about an organism—what it does and to whom it is most closely related. Using only the taxonomic classifications, we can tell that house cats eat meat (order Carnivora), have warm blood and fur (class Mammalia), have spines (phylum Chordata), and are animals (kingdom Animalia). We can also tell that peaches (family Rosaceae) are closely related to roses.

Linnaean taxonomy remains helpful for identifying species. Gardeners, bird watchers, and naturalists commonly use the Latin binomial (genus and species). Zoologists, botanists, and mycologists (who study animals, plants, and fungi, respectively) also use this system daily. At the same time, some evolutionary biologists and biologists who study one-celled organisms have found that the Linnaean system presents more problems than it solves. The most fundamental problem has to do with species. As long as you are studying mammals, the idea of species makes sense. Two members of the same species can reproduce and have viable offspring. Members of different species cannot. Many plants and animals follow this simple pattern and fit well into the system. Most one-celled organisms, on the other hand, reproduce by division. A single cell divides in two. No compatibility required. So what constitutes a species of bacteria?

Some molecular biologists have suggested that species should encompass any group with 97 percent genetic identity. Any group that has less than 3 percent difference in their genes should be considered a single species. This would lump chimpanzees and humans together and make the term "species" not specific enough for our purposes. On the flip side, members of the species *Escherichia coli,* common bacteria in human stomachs, can be as much as 40 percent different. Once again, we see that a unit can be useful in daily life—and even in scientific exploration—but fail to be consistent in certain cases. We must remember that genus, species, and all the other taxonomic categories remain popular convention, but do not have any reality in nature. They are human categories that we can use to clump organisms together and make them easier to discuss.[4]

I have said earlier that one of the great wonders of astrobiology is the idea that all terran life represents a single example of life in space. Phylogenetics creates a great tree of life that connects all living organisms (and all extinct organisms) in one giant family of relationships. For at least the last thirty years, biologists have tried to readjust taxonomy so that the hierarchy of names represents real branches on that tree. In large part they have succeeded. In trying to classify life and understand its complexities, we have discovered that our initial assumptions were inadequate. Genus and species, though often useful categories, are human constructions. For everything we have lost, however, we have gained something new—an appreciation for the fundamental interconnectedness of life on Earth.

Communities of Common Interest or Location

Earlier I introduced the idea that selection acts to favor individuals, or perhaps functional units. It undoubtedly does. Selection can also foster cooperation. From small to large, these cooperative groups, or communities, interact with and change their environment, sometimes even influencing the geological and chemical cycles of Earth. Ecology, the study of relationships among organisms and between organisms and their environment, looks at the next step on the ladder of complexity.

The word "symbiosis" comes from the Greek words meaning "to live together." The word comes with a certain amount of confusion because it can mean simply living together or living together for mutual benefit. There are many kinds of symbiosis. Both organisms may benefit (mutualism), one organism benefits without causing any harm to the other (commensalism), or one organism benefits at the expense of the other (parasitism). Organisms working together can achieve things that the individuals could not. Tubeworms living at the bottom of the ocean *(Riftia pachyptila)* exist only because tiny sulfur bacteria live within their tissues.[5] The tubeworm could not exist without the bacteria that generate energy from reduced sulfur compounds coming out of deep-sea vents (chemolithotrophy). The bacteria benefit from the stability provided by the tubeworms' size and attachment to the seafloor. Alternatively, parasitic bacteria like *Ehrlichia* act as parasites within insects, limiting their reproduction and draining resources. Relationships can be more complicated than they appear, however, as two interrelated species evolve together. Mitochondria, the eukaryotic organelles responsible for processing sugars, are close cousins to several intracellular parasites. Ancient parasites may have become trapped within their host and become permanent members of their biological environment. The lines between mutualism, commensalism, and parasitism are often blurry. A symbiote may provide a benefit we do not see—or come with a hidden cost. Symbioses represent new living structures and new functionalities—a "higher level" of life than simple organisms. Often this higher level simply means that one organism has discovered a new niche, but sometimes the two working together can do things that neither one could do alone.

Ecosystems represent more extensive relationships among organisms. An ecosystem encompasses all the organisms in an area—from primary producers, which convert light into energy, to the animals that eat them and the bacteria that break down the animals when they die. Ecologists assign ecosystems somewhat arbitrarily in an attempt to bite off a chewable portion of terran life. The web of interactions across the face of the planet contains far too many processes for anyone to track all of them at once. An ecosystem, like a desert or a rain forest, has more discrete boundaries and more easily tracked processes. Ecologists frequently attempt to trace the path of resources such as carbon, nitrogen, or water from one organism to another. This allows them to see how the organisms relate, compete, and coexist. Symbiosis implies a very close relationship, often with two organisms in constant physical contact. Many other relationships involve only occasional contact or the use of intermediaries. Nonetheless, a whole chain of interactions will be necessary for the survival of most species. Humans, for instance, rely heavily on the growth of soybeans. Soybeans become feed for livestock later consumed by humans. So humans depend on soybeans and animals as a source of carbon and energy. We also depend on the metabolism of cyanobacteria in the oceans. The cyanobacteria consume carbon dioxide and produce oxygen, keeping the air breathable for all animals. Tracing the passage of water, nutrients, and energy around the planet has led ecologists to see all organisms as fundamentally interconnected.

Biospheres

Biospheres are self-contained ecosystems, requiring no input but energy. To date only "biosphere 1," Earth, is known to exist. Attempts to construct enclosed systems with humans inside include BIOS-3, a 315 m^2 habitat in Krasnoyarsk, Siberia, and Biosphere 2, a three-acre (~12,000 m^2) terrarium in Oracle, Arizona. The first survived a 180-day run with three people (1972–1973) and the second a two-year run with eight people (1991–1993). The two experiments provided important information about the nutritional and energy budgets of humans and other organisms. This information will inform plans for future space missions, particularly the construction of habitats on the Moon or Mars. Unfortunately, all variables could not be controlled and long-term self-sufficiency is not currently possible.

Vladimir Vernadsky introduced the term "biosphere" in 1929 to include all living organic material on the planet, contrasting it with the atmosphere, the envelope of gas around the planet; the hydrosphere, the liquid water on the planet; and the geosphere, the rocky core and crust of the planet. The more we learn about the biosphere, the more we realize how intimately connected it is to the other realms. Oxygen produced by cyanobacteria and plants forms the ozone layer, which has a profound effect on atmospheric and hydrospheric chemistry. Deposits of organic materials from dying organisms change the chemistry of the upper crust. The tectonic exchange of material so central to our concept of Earth's uniqueness turns out to be paralleled in the circulation of nutrients in the biosphere.

A Common Narrative

Looking at scale allows us to see many of the important interactions that shape life. Astrobiology in particular observes the boundaries between levels and tries to piece together a single, comprehensive narrative. As it turns out, however, there is not one story of life but many; it can be told from a number of perspectives. It can be told chronologically, starting at the beginning of the universe and continuing until the present. It can be told chemically, looking at the input of energy from the Sun—how the energy gets trapped in the biosphere through the propagation and reproduction of organisms. It can also be told hierarchically, starting with quarks and watching the accumulation of more and more complex systems for storing information and performing tasks.

Let us return for a moment to a fourth way of telling the story. Recall the issues of evolution, reductionism, and emergence raised in Chapter 2. Charles R. Marshall, professor of biology and geology at Harvard University, makes a compelling argument about scale and complexity. He says that some things—such as watches found on beaches—really do show evidence of an intelligent designer. Each of the pieces interacts with other parts of the mechanism in exactly one way. Different cogs and toggles fit together precisely only because each pair was individually calibrated. This represents top-down organization—organization imposed from above. Life, on the other hand, shows evidence of building from the bottom up. At each scale, we see a very small number of building blocks that can be assembled in nearly infinite ways. The blocks fit together at larger and smaller scales in a way far subtler, and yet with more diversity than the pieces of the watch. Life explores a vast array of possibilities, fitting things together not optimally but successfully.

The amazing stackability of units from the smallest quark to the terrean biosphere tells us something important about life. The radical interrelationship of building blocks at every conceivable scale looks like bottom-up organization. Energy poured into a system of units causes them to assemble and disassemble in numerous ways until they stumble into a configuration that perpetuates itself and holds the energy in place. Think back to the watery volcano metaphor.

Earth sits at the perfect place in the Solar System. Energy streams in from the Sun at a number of wavelengths. Ozone in the atmosphere blocks the more powerful radiation, leaving visible light the most prominent source of energy on the surface. Atoms generated in the death of giant stars tumble through the liquid water, combining and recombining to form larger and larger molecules. Would this be enough to initiate life? We don't really know. We cannot replicate the next step, though we continue to investigate how it might have occurred. Once self-replicating systems arose, energy trapped in the system ensures more and more complex biochemistry. The interaction of heredity, variation, and selection leads to an increasingly broad range of biochemical possibilities.

Once it starts, it must continue, as long as energy continues pouring in. We can imagine the development of more and more complex structures, with proteins catalyzing reactions and membranes holding reagents together—or apart—until the opportune time. Organisms as we know them develop as cells with internal structure and ways of interacting with the environment. Some one-celled organisms fail to reproduce properly, leading to many cells attached together, which in some rare circumstances proved advantageous. Multicellular life was born. Organisms learned to take advantage of other organisms, resulting in symbioses and communities. The process builds.

The biosphere, of course, must have existed from the very beginning. As soon as life exists, a biosphere exists. As time passed, it became more and more complex, storing energy not only in the seas, but on land, in the air, and deep below the surface. Some interactions between organisms ceased—some organisms, species, whole kingdoms went extinct—but the remainder stayed tied together in a web of mutual dependence and interaction. New resources were discovered and exploited to develop new life until the entire surface of the planet formed one giant ecosystem.

One of the greatest benefits of astrobiology has been the chance to question fundamental ideas about life. Dwelling in the borderlands of so many different scientific disciplines, astrobiology forces us to ask if our definitions are clear, consistent, and useful. How do my ideas shape the way I think about life? What biases do I have and how do they affect my ability to observe the cosmos around me?

Scientists have the task of prying into the structure of the universe, but each of us must come to our own peace about what it all means. Astrobiology forces us to recognize the very real links between organisms and their environment. In light of this, each of us must do the difficult and sometimes tedious work of looking at all the messy details of life as it actually happens. Often the results are surprising.

15

Molecules

How do elements come together to form the structure of life as we know it? The first step is the construction of molecules. Almost every structure in your body was built out of one of only four types of biological molecules—carbohydrates, proteins, nucleic acids, and lipids. Each kind of "biomolecule" ranges in size from a few atoms joined together to huge strings that can involve thousands of atoms. In this chapter, I introduce some of the immense variety of biological molecules found in terran life.

Each section begins with a discussion of how to get from atoms to simple molecules. Six elements make up 95 percent of all organic matter—carbon, hydrogen, nitrogen, oxygen, phosphorus, and sulfur. Using carbon chains as a backbone, life adds functional groups to make the molecules useful for biological work. The first three types of biomolecule, carbohydrates, proteins, and nucleic acids, each come in a number of sizes. Small units, called monomers, join together to form long chains or polymers. So carbohydrates are built out of sugars, proteins out of amino acids, and nucleic acids out of nucleotides. Lipids can also come in multiple unit strings, but the process is a little more complicated.

After covering the atomic foundations of the biomolecules, I mention some of the most significant examples. We encounter each type of biomolecule in our daily lives and can learn to recognize the important biochemistry involved.

Sugars and Carbohydrates

Carbohydrates make up the first category of biomolecules. They are composed of one part carbon, one part oxygen, and two parts hydrogen—or something close to that ratio, hence the name "carbohydrate." Carbohydrates should be familiar to most readers because we use them for fuel. Animals have a metabolism that burns sugar, the building blocks of carbohydrates, in a process called respiration. Carbohydrates come in one of three forms: monosaccharides (with one simple sugar), oligosaccharides (with just a few simple sugars), and polysaccharides (with many sugars strung together).

Monosaccharides

Simple sugars, or monosaccharides, consist of a single unit: a string of carbons, each attached to a single oxygen, with hydrogen filling in the gaps. The chain can be laid out straight or form a loop. I have drawn the most common biological sugar, glucose, in Figure 15.1. To the left, you can see the arrangement of all the atoms. Each carbon forms four bonds, each oxygen forms two bonds, and each hydrogen, one. Because biochemists frequently want to draw organic molecules like this one, simpler drawing conventions have arisen. The carbons can be represented simply as intersections between lines, and hydrogens bound to carbons disappear completely. We assume that any extra bonds available from a carbon will be attached to a hydrogen. The upper right drawing shows the simplified molecular diagram. The carbon at the top will frequently react with a carbon near the bottom to form a ring structure. A ring glucose appears to the lower right. Take a moment to convince yourself that all three diagrams show the same assembly of atoms.

Diagrams on paper can only cover two dimensions, but when a car-

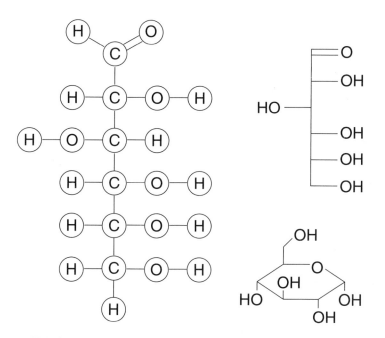

15.1 Glucose. The sugar glucose appears here as a molecular diagram (left) and as a stick diagram (right). Glucose occurs in both linear form (upper right) and ring form (lower right).

bon makes four bonds, it makes them in a tetrahedral configuration. Imagine a pyramid with three sides (plus one triangular side facing down). This is a tetrahedron. Now imagine that all four faces are equilateral triangles with three equal sides. If you placed a carbon atom in the middle, its four bonds would point out along the four points of the tetrahedron.

If the four bonds connect to four different types of atom or four different portions of the molecule, then the carbon is said to be chiral. It does not resemble its own mirror image. One of the most interesting features of glucose appears in Figure 15.1. If you look closely at the illustration on the left, the third carbon from the top has oxygen on the left instead of the right. Chemical diagrams contain an added level of information regarding the chirality of molecules. A molecule can be right- or left-handed. Chemists usually make stick diagrams so that they would be accurate if the carbons were arching three-dimensionally back into the page at the top and the bottom.

Life cares about chirality. For our purposes, the relevant detail of chirality in sugars is that the orientation of the hydroxide ($-OH$) groups matters. The sugar in Figure 15.1 corresponds to D-glucose—its mirror image would be L-glucose. Tests using polarized light can differentiate right- and left-handed versions of a molecule. The "D" and "L" help us distinguish between the two.[1] For unknown reasons, terrean life runs on the former and cannot process the latter. This may simply be a legacy of the first organisms, which used D-glucose. Whatever the reason, protein shape dictates the use of only one type of glucose.

To make matters even more complicated, sugars can form rings in one of two ways. The newly formed bond creates a new chiral center, the arrangement of which depends on how the new bond is made. Figure 15.2 shows the two different possible rings—with the rightmost hydroxide above the plane or below. Chemists distinguish between the two types of ring by calling the first alpha (α) and the second beta (β). In solution a monosaccharide can change freely between the linear conformation and both ring structures; once sugars join together, however, they can become fixed in either the α or β conformation. So, the full name of a sugar, including chiral designations, includes an English letter (D or L) and a Greek letter (α or β). The important thing to remember is that the chirality of molecules can impact how they react chemically. Biochemical reactions, in particular, can be very sensitive to the handedness of molecules. Having noted the importance of chirality in carbohydrates, I will stick to common names from here on out.

Glucose is important, but several other monosaccharides warrant a mention as well. Ribose is a pentose, meaning it has five carbons. It forms a five-atom ring with one carbon hanging off (glucose makes a six-carbon ring). Ribose helps to form the backbone of nucleic acids (see Figure 15.7, below). Common hexoses (six-carbon sugars) include glucose, fructose,

LIFE IN SPACE

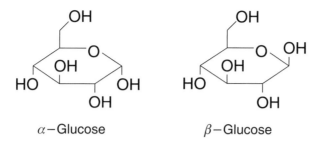

α−Glucose   β−Glucose

15.2 Chiral rings. The figure shows two conformations of D-glucose. The two names reflect the different arrangement of side groups on the rightmost carbon.

mannose, and galactose. Fructose, or fruit sugar, is common as a monosaccharide in honey and fruits. Mannose and galactose are important parts of larger carbohydrates.

Oligosaccharides

Short-chain sugars, or oligosaccharides, consist of two or more simple sugars joined together by glycosidic linkages. The linkage occurs when one of the OH groups in one sugar runs off with a hydrogen from an OH in another sugar. A dehydration reaction produces water and the two sugars become linked (Figure 15.3; see also the dehydration reaction in Figure 6.2). Meanwhile, the sugars have been joined with an ether bond (-C-O-C-). Sucrose, or table sugar, combines glucose and fructose in this fashion. Lactose, the sugar in milk, combines galactose with glucose (Figure 15.3). As a side note, lactose intolerance comes from a deficiency in lactase, the enzyme that separates the glucose from the galactose. You cannot digest lactose without this step, and gas forms as intestinal bacteria digest it for you.

Polysaccharides

Organisms use glycosidic linkages to string together immense chains, thousands of sugars long. Polysaccharides, made up of many sugars, make storage and structural molecules. Two common examples are starch and

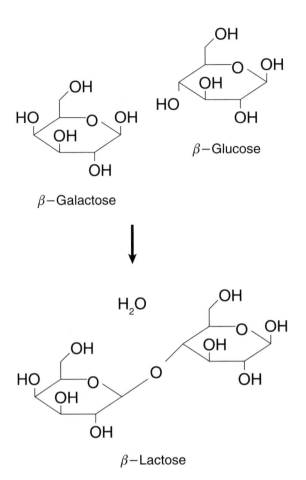

OH

β–Glucose

OH

HO

β–Galactose

H₂O

OH

HO

OH

O

β–Lactose

15.3 A glycosidic linkage. β-galactose and β-glucose join together to form β-lactose (milk sugar).

cellulose, both composed of glucose. The difference between them depends on the orientation of the units when they join together.[2] The two chains have remarkably different structural and biochemical properties. Polysaccharides play an enormous role in terrean biochemistry, but I will just mention a few of the more important polysaccharides here.

Starch is an essential polysaccharide, at least from the perspective of animals. Plants use starch as a storage device. They pack glucose units together like strings of sausage, pulling off sections and eating them as needed. Through respiration, glucose can be broken down, making chemical energy available for work. Rice, corn, wheat, and potatoes form the basis of the human diet because each contains large quantities of starch, packed up for future use by the plant and harvested by humans.

Cellulose occurs more abundantly than any other natural polymer. This glucose chain makes up the main structural component of plants. Cellulose forms the cell wall and is the principal component of wood.[3] Most animals cannot digest cellulose (although humans have come to appreciate the value of "fiber," which passes through and cleans out the digestive system). Only one group of animals has developed the ability to digest cellulose, but you wouldn't call them close cousins. Tunicates (sea squirts) share less in common with humans than all vertebrate species, including fish. Ruminants, hooved mammals with four stomachs that chew their cud, cannot actually digest cellulose, but they maintain colonies of bacteria and other one-celled organisms within their stomachs.[4] The symbiosis benefits both the animals, which take advantage of a new food source, and the bacteria, which get transportation, protection, and a regular supply of food.

The structure of cellulose provides both flexibility and strength, making it ideal for clothing when not mixed with lignin or certain other materials. Cotton fibers have little in them except cellulose, making them ideal for weaving soft fabrics. An artificial alternative, rayon, represents chemically purified cellulose. Cellulose has also been central to the papermaking industry, forming the rigid structure of ancient papyrus and the more versatile properties that come from wood pulp.

Chitin closely resembles cellulose. It has an almost identical string of glucose molecules, except each has been modified. A side chain has been added to the second carbon, making it a glucosamine.[5] Chitin makes up the structural component of fungi and the exoskeletons of arthropods, which include insects and crustaceans as well as centipedes, millipedes, scorpions, ticks, and spiders.

One final polysaccharide merits mention here. Peptidoglycan forms the cell walls of bacteria. Bacterial cells face a number of structural challenges. Although they do not build skyscrapers like the redwoods, they do have to keep their internal and external chemistry separate, sometimes under extreme conditions. The cell wall resembles a fabric, with polysaccharide strands passing in one direction and strings of amino acids crossing in the other direction. Linked together, the chains make a very solid scaffolding for the cell membrane.

Any number of sugars can be strung together in a variety of ways to form a variety of molecules. The carbohydrates that are created serve as essential energy storage cells and structural units for a wide array of organisms. The importance of glucose, in particular, cannot be underestimated. This simple sugar contributes to the formation of starch, cellulose, chitin, and peptidoglycan—four ubiquitous, important, and easily recognized biomolecules.

Amino Acids and Proteins

The second category of biomolecules makes proteins. Unlike carbohydrates, proteins follow a very limited code. Sugars come in hundreds of forms, but almost every organism on Earth uses a limited repertoire of only twenty amino acids. These twenty units can be used to assemble a nearly infinite number of peptides and proteins. The proteins, meanwhile, serve in a tremendous number of functions. They can do everything from carrying information to packaging DNA; from breaking apart starches to assembling lipid membranes. Proteins do the work of life at the molecular level. So we see again the pattern of a few building blocks assembled in a huge number of ways.

Amino Acids and Oligopeptides

Amino acids are practically defined by their ability to link together. Imagine amino acids in a train, with each acid as a single boxcar. The boxcars all serve different purposes, but they hook together in the same way. Amino

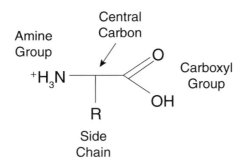

Amine Group

Central Carbon

^+H_3N

Carboxyl Group

O

OH

R

Side Chain

15.4 An amino acid.

acids get their name from the two functional groups that allow them to connect. Figure 15.4 shows the structure of a generic amino acid. At one end of the molecule we find an amine group ($-NH_2$ or $-NH_3^+$). At the other end, we find a carboxyl group ($-CO_2H$ or $-CO_2^-$). In between lies a single carbon atom attached to each as well as to a hydrogen and a side chain that defines the specific amino acid. Using the train metaphor, the central carbon acts as a flatbed car with couplings at both ends, the amine and carboxyl groups. The side chain makes the cargo box. Any number of things can be placed on the train, as long as they fit on the bed.

How do the boxcars link together? Much like sugars, amino acids can be joined with a dehydration reaction. Hydrogen and hydroxyl come together to form water when the two molecules join. Unlike glycosidic linkages, however, amino acids form a bond with both a carbon and a nitrogen (-C-O-N-C-). Biochemists call this a peptide bond. From peptide bonds we get the names oligopeptide (short chain of amino acids) and polypeptide (long chain of amino acids). Figure 15.5 shows a peptide bond forming.

Amino acids come in a wide variety of shapes and sizes. Some are basic; some acidic. One has nothing more than a hydrogen as a side chain (glycine); four have complicated ring structures. All amino acids (except glycine) are chiral; their central carbon attaches to four different groups. Naturally occurring polypeptides usually have the L configuration, demonstrating another chiral bias in terran life.[6]

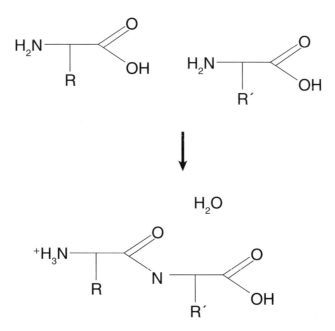

15.5 A peptide bond. Two amino acids link together to form a dipeptide. (Note that R and R′ in the figure represent generic molecules. A huge number of possible Rs and R′s exist.)

Terrean life requires twenty amino acids. Although similar molecules occur in biochemical processes, only the standard twenty appear in proteins. Many organisms produce their own amino acids from scratch and, in rare cases, modify them for specific purposes.[7] Other species suffer from an inability to produce certain amino acids and must acquire them through food. Humans cannot produce eight such "essential" amino acids. They are essential because we need regular supplements to make our proteins. Our nutritional requirement for protein rests on these essential amino acids and a general need for nitrogen.

Polypeptides and Proteins

The twenty amino acids can be strung together in myriad ways to perform different functions in the cell. Some polypeptides run messages between different parts of the body. Others catalyze reactions, speeding up chemical processes, or act as gateways, selectively passing some molecules through membranes. Polypeptides form the substance of muscle fibers, contracting to allow motion. They even act as structural components.

Functional polypeptides range in length from very short to very long. The peptide hormone vasopressin, which acts to conserve water in the human body, has only nine amino acids. The structural protein collagen, found in cartilage, can be over 1,800 amino acids long.

The versatility of the different amino acids allows them to make connections between different places on the chain, changing the overall shape of the protein. A small set of very common shapes form on the basis of hydrogen bonds—weak attachments between positively and negatively charged side chains. These connections can wrap up the molecule into helices, creating a strong, rigid structure. And, if the helix has all the polar groups on the inside, it will avoid water and fit well in a membrane. Hydrogen bonds can also make pleated sheets that give a molecule strength and flexibility. The protein fibroin has many such pleated sheets and forms the main component in silk.

Even more complex folding occurs within many polypeptides. Helices will align or sheets will stack up. The number of possible structures is vast, but I would like to mention two types of structure that have important implications for terran life.

Fibrous proteins contain long strands useful in constructing fibers and surfaces. These proteins play an important role in the creation of tissues and polymers. Keratin, for instance, resembles twine. Helices entwine to form a coiled coil. Multiple polypeptide chains can be spliced together to form one very strong rope. Keratin is the structural component of hair and nails in humans; fur, horns, claws, and hooves in other mammals; feathers, beaks, and claws in birds; and scales, shells, and claws in turtles, dinosaurs, and other reptiles. Collagen also utilizes a coiled coil pattern, though not one as convoluted as keratin. In collagen, three polypeptide strands weave together to form an immensely strong fiber. The rigidity and tensile strength make fast movement possible for animals. Collagen makes up the structural component of connective tissue. It is essential to bones, blood vessels, cartilage, ligaments, teeth, and skin. It represents about one-quarter of all protein in mammals.

Globular proteins, though less well characterized, make up the majority of terran proteins. In solution, they ball up to form a rough sphere or glob—hence the name. A number of globular proteins associate with a complicated carbon ring structure that can hold a single oxygen molecule. Biochemists call the ring a heme and the protein a heme protein. In mammals, hemoglobin carries oxygen in the bloodstream, myoglobin in the muscles, and neuroglobin in the spinal fluids. All three associate with heme. Some plants use leghemoglobin to regulate oxygen levels around nitrogen-fixing bacteria. Another variety of hemes common to all organisms, cytochromes, transport electrons. These are only the most common examples of a heme, a protein mechanism important to life.

Photosynthetic reaction-center proteins catch high-energy photons in a carbon ring structure very similar to heme. The prosthetic group is called chlorophyll.[8] In photosynthetic bacteria, a photon excites the chlorophyll, charging an electron that can then be used to reduce nearby molecules. In this way, light energy gets converted to chemical energy, powering the biosphere.

Already, we can see that proteins have a profound impact on terran life. Their structure may seem complex at first, but careful investigation shows that life has a limited number of options at every stage of the process. First, a number of amino acids must be strung together in sequence. Second, simple interactions between hydrogen molecules on those amino acids shape portions of the polypeptide into sheets and helices. Third, interactions with the environment and local molecules (such as heme) shape whole chains into ropes, or globs, or other formations that allow them to do biological work. Finally, multiple polypeptide chains can bind together to do even more. Any chain with more than a few amino acids qualifies as a polypeptide, no matter where it occurs or what it does. A polypeptide or set of polypeptides with a biological purpose constitutes a protein.

LIFE IN SPACE

Enzymes

Proteins serve numerous functions. They operate as signaling molecules and hormones (e.g., vasopressin). They form structural components (e.g., collagen). They carry electrons and oxygen (e.g., cytochrome). They even convert light into chemical energy (photosynthetic reaction centers). I have yet to mention the most famous ability of proteins, however. They can speed up chemical reactions. Proteins that catalyze reactions, making them run faster or more smoothly or at a lower temperature, are enzymes.

The most common enzyme on Earth goes by the name "rubisco," short for ribulose bisphosphate carboxylase/oxygenase. Rubisco adds carbon dioxide to a five-carbon sugar and makes two three-carbon sugars—the first step in the metabolic pathway by which plants fix carbon. Rubisco brings carbon into the biosphere. Inorganic carbon dioxide becomes organic sugars that can later be assembled into any number of molecules. More than any other biological process, this one reaction is responsible for the use of carbon in life.

Proteins do the work of life. Animals depend on proteins for structure, movement, and circulation. Keratin, present in fur, feathers and scales, forms the tough, fibrous material that defends animals from the outside world. Collagen makes up ligaments and bones, and hemoglobin and myoglobin transport oxygen to individual cells. Beyond this, proteins do chemical work. Enzymes shepherd chemical reactions at every point in terrean metabolism, speeding up reactions, regulating concentrations of chemicals, and even helping to process information.

Nucleotides and Nucleic Acids

The third category of organic molecules consists of nucleotides and nucleic acids, the information keepers. Even more limited than amino acids, only four nucleotides make up the complete range of monomers, though one of those nucleotides takes a different shape in the two different nucleic acids. Deoxyribonucleic acid (DNA) acts as the central information stor-

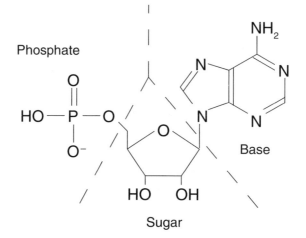

15.6 A nucleotide. The figure shows the three major parts of a nucleotide, the basic unit forming nucleic acids.

age molecule in almost all terran organisms. Ribonucleic acid (RNA) acts as an intermediary, conveying information from the DNA to the ribosomes, helping to construct proteins, and, in the absence of DNA, storing information. The nomenclature for nucleic acids can be a bit overwhelming at first, but also highly informative. The basic molecules of DNA contribute to ATP, NADH, and other fundamental biomolecules.

Nucleotides

Every nucleotide monomer consists of three parts—a phosphate ion, a simple five-carbon sugar, and a base. Figure 15.6 shows how the three parts fit together. All nucleotides have the same phosphate ion (HPO_4^{2-}). All nucleotides have one of two sugars. Figure 15.7 shows the two different sugars, ribose and deoxyribose, which form RNA and DNA respectively. As you can see from the diagram, ribose (on the left) is a pentose; deoxyribose is the same sugar with one OH removed. The phosphate bonds to the sugar and the sugar bonds to the base.

In DNA, the deoxyribose attaches to one of four bases: adenine, cytosine, guanine, or thymine. All four appear in Figure 15.8. Biochemists

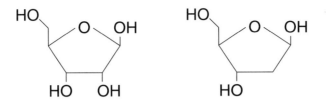

15.7 Nucleic acid sugars. Ribose (left) and deoxyribose (right).

commonly use the one-letter abbreviations A, C, G, and T, respectively. Two of the bases, C and T, are called pyrimidines. Pyrimidines form with a single, six-atom ring composed of four carbons and two nitrogens. The purines, A and G, have two joined rings, with five carbons and four nitrogens. In RNA, the ribose may attach to adenine, cytosine, or guanine,

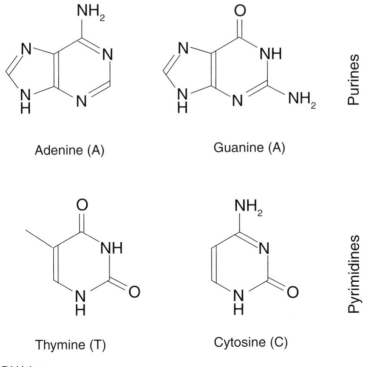

Adenine (A)

Guanine (A)

Purines

Thymine (T)

Cytosine (C)

Pyrimidines

15.8 DNA bases.

MOLECULES

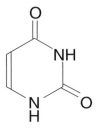

Uracil (U)

15.9 Uracil. RNA shares bases A, C, and G with DNA, but replaces T with U.

but the fourth base, thymine, has been replaced by uracil (U) (Figure 15.9).

Starting with one ion, two types of sugar, and five different bases, we can assemble a number of interesting molecules (Figure 15.10). Biochemists refer to a sugar plus a base as a nucleoside.[9] In a nucleoside, an OH group from the sugar has run off with a hydrogen from the base. The first carbon of the sugar becomes linked to one of the nitrogens in the base. Add a phosphate ion, and it becomes a nucleotide rather than a nucleoside. In nucleotides, the phosphate group connects to the sugar by way of the one carbon outside the ring. Nucleotides have one, two, or three phosphate ions attached in sequence. Adenosine, for example, becomes adenosine monophosphate (AMP, 1 ion), adenosine diphosphate (ADP, 2 ions), or adenosine triphosphate (ATP, 3 ions). AMP, ADP, and ATP are all nucleotides.

All of this nomenclature may seem a bit excessive, but I wanted to introduce it because of the extreme importance of ATP. ATP functions as the primary unit of energy in all cells. By attaching additional ions to AMP, the cell stores energy in ATP or ADP. That energy can later be released by popping off an ion. The breakdown of a single glucose molecule, for instance, can yield 38 ATP equivalents.[10] Glucose formation, on the other hand, requires 54 ATP equivalents.[11] This gives further proof that entropy increases, even in biology; more energy goes into the making of glucose than comes from the destruction of glucose.[12]

LIFE IN SPACE

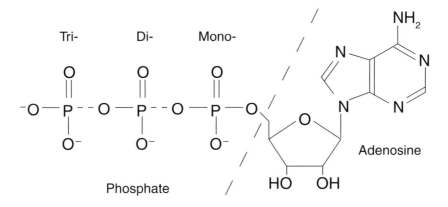

Tri-   Di-   Mono-

Phosphate

Adenosine

15.10 Nucleotide names. In the figure we see a nucleoside, adenosine, on the right. It attaches to one phosphate to make AMP, two phosphates to make ADP, or three phosphates
to make ATP. In all three cases, a nucleotide has formed.

Dinucleotides

Nucleotides can join together in two very important ways (Figure 15.11).
First, polymers—nucleic acids—form by attaching nucleotides end to
end. Second, base pairing, the ability of nucleotides to line up with their
counterparts (A-T or A-U, C-G), allows two nucleic acids to align in a
way that makes information storage easy and efficient, as I explain below.

Nucleic acids form when an organism strings together monophosphate
nucleotides. Each phosphate ion has a double attachment—with one oxygen linked to each of two sugar molecules. A single chain can contain tens
of millions of nucleotides. Every known organism contains a genetic blueprint, with instructions for making new organisms and maintaining a metabolism. This information resides within one or more nucleotide chains,
either DNA or RNA. The term "nucleic acid" can refer either to a single
polymer or to a collection of polymers. Nucleic acids, plural, usually include both DNA and RNA.

One of the most fascinating and useful aspects of nucleic acids comes
from base pairing. Base pairing refers to the ability of purine-pyrimidine
pairs to form hydrogen bonds, giving them a weak but significant attach-

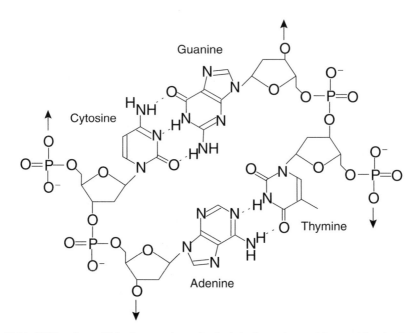

Guanine

Cytosine

Thymine

Adenine

15.11 DNA polymer. This diagram shows both chain formation and base pairing. In chain formation, nucleic acids form from alternating sugar (deoxyribose) and phosphate molecules. In base pairing, the two strands stick together because of the hydrogen bonds between adenine and thymine and between guanine and cytosine.

ment. In DNA, thymine bonds with adenine, and cytosine bonds with guanine. In RNA, uracil takes the place of thymine, forming a bond with adenine. Two polymers can be joined from end to end, across millions of base pairs, by these hydrogen bonds. The paired strands form a characteristic double helix as the chains wrap around each other, usually turning 360° every ten bases.

The discovery of the structure of nucleic acids was reported by James Watson and Francis Crick, with the help of Rosalind Franklin and Maurice Wilkins, in 1953. Despite years of research and complex modeling, the final structure was so simple that it could be described in less than two pages. Watson and Crick's paper, describing one of the most important discoveries in biology, remains a paramount example of literary brevity and an elegant theory.[13]

Deoxyribonucleic Acid (DNA)

DNA serves to store information within the cell. Cell biologists call the large unit of DNA used by an organism to store basic information a chromosome. In eukaryotes, chromosomes form as extremely long nucleotide chains, condensed and packed with numerous proteins. The proteins stow most of the DNA but allow discrete portions to be unfurled at need. Given the tremendous volume of information, packing can be an important concern. Estimates place the length of a single nucleotide chain, a single chromosome, in humans between 2 and 3 cm. That entire length needs to be stored in a packet that fits neatly into a cell 2 to 3 μm in diameter and requires at least a ten-thousandfold contraction in size. These eukaryotic chromosomes all reside within the nucleus.

Prokaryotes store their DNA in rings. The circular chromosomes have no membrane-bound nucleus but are often confined to a region called the nucleolus. Microbiologists are increasingly aware of tiny linear and circular pieces of DNA that operate independently of chromosomes (see Chapter 18).

Genes—How Information Is Stored

DNA forms its own hierarchy of scale. At the most fundamental level, DNA constitutes a series of bases or base pairs. Above the nucleotide level, we find codons, genes, and genomes. It can be difficult to talk about any one level alone, so I will introduce them all briefly and then try to tie them together. A codon is a set of three nucleotides used for coding proteins, whereas a gene is a functional unit of inheritance, usually hundreds of nucleotides in length. A genome is the complete set of nucleotides within chromosomes for a given organism.

Estimates place the entire human genome somewhere near 3 billion base pairs in length. The smallest known genome comes from *Mycoplasma genitalium,* with only 580,076 bases. If we count viruses, the number becomes even smaller; the hepatitis B virus contains a genome about 3,000 bases in length. It would be tempting to assume that larger genomes correspond to larger or more structurally complex organisms, but this is not the case. The largest genome on record comes from a one-celled protist, *Amoeba dubia,* with approximately 670 billion base pairs.[14]

At present, no clear relation between genome size and functional or structural complexity exists. Some of the extra bases contribute to metabolic diversity, the ability to perform various chemical processes. Other regions may be useful in recombination, increasing variation within a species. Some portions of the DNA are simply opportunistic information that causes the cell to replicate it over and over again with no other known function.

The term "gene" was coined by Gregor Mendel (1822–1884), an Augustinian friar, botanist, and the father of modern genetics. Mendel first described discrete units of inheritance in the late nineteenth century, but was unaware of how these units worked physically. It would take the molecular biology of Watson and Crick nearly one hundred years later to make the connection between inheritance and nucleic acids meaningful.

Modern biologists tend to see genes as discrete stretches of nucleotides responsible for a single function in an organism. Genes that code for proteins are the easiest to understand. A sequence of nucleotides can be read like a coded message. Every three nucleotides correspond to one amino acid according to a pattern used by almost every organism on the planet. Geneticists call the three base-pair units codons. In this way, the cell stores the sequence of a protein within the DNA. Every gene has additional information located before and after the coding region, stretches of DNA that help proteins locate the coding region and know when to copy it out. In eukaryotes, the genes may also include introns, portions of DNA within the coding region that need to be clipped out before the message makes sense. Prokaryotic DNA lacks introns.

Some genes do not code for proteins, though these genes tend to be less well understood. Chromosomes contain many regulatory regions as well as regions that code for certain types of RNA, discussed below. Much DNA has no known purpose, yet appears essential for the health of the

LIFE IN SPACE

organism. Still more DNA has no apparent function at all. If a gene simply means a unit of inheritance, then many noncoding genes must exist. DNA has a complex relationship with structural proteins and interacts chemically with its surroundings. The coding regions are the easiest to identify and name; therefore, almost all presently named genes code either for proteins or RNA.

Known genes range from very short (tens of bases) to very long (thousands of bases) and may be spread out over a chromosome. Some genes need to be pieced together from multiple locations before they can be translated into proteins. By current estimates, humans possess between 20,000 and 25,000 genes. *Mycoplasma genitalium* has about 400 and the hepatitis B virus has only 4. *Amoeba dubia* is not yet well enough understood to estimate.

Ribonucleic Acid (RNA)

The function of nucleic acids is not limited to information storage. RNA has a far more active role than DNA. It serves as an intermediary between DNA and the rest of the cell, but also catalyzes reactions like a protein. A protein enzyme called RNA transcriptase makes a reverse image of DNA by lining up complementary bases and creating a strand of messenger RNA (mRNA). Messenger RNA travels from the site of the gene (through the nuclear membrane in eukaryotes) to the ribosomes, organelles that construct protein. Transfer RNA (tRNA) brings amino acids to the ribosome, where they can be strung together. Every tRNA comes with two binding sites. At one end the tRNA links up with the amino acid. The other end matches up with a codon in the mRNA. Ribosomes create peptide bonds between the amino acids, creating a new polypeptide.

Each cell has tens of thousands of ribosomes, each with a minimum size of 50 nm in diameter. Ribosomes represent a complex association of nucleic acid (35 percent) and protein (65 percent). All ribosomes contain three nucleic acid chains made up of ribosomal RNA (rRNA). The small subunit (SSU rRNA) consists of a single chain ~1,500–2,000 nucleotides in length. The large subunit has two chains in prokaryotes (~3,000 + 120 bases) and three chains in eukaryotes (~5,000 + 160 + 120 bases). Ribosomes also contain over fifty proteins.[15] Eukaryotes have larger ribosomes than prokaryotes, and plants, animals, and fungi have larger ribosomes than other eukaryotes.

The Central Dogma

I've condensed a great deal of information into a very small space. The whole process, however, need not be difficult to grasp. The flow of information in a cell can be neatly summed up in a theory called the central dogma of molecular biology. In short, the central dogma states that DNA codes for RNA and RNA codes for proteins. A more detailed description goes like this: Genetic information in DNA comes only from DNA. (You can see how this might have interesting implications for astrobiology and the origin of life.) All genes have been copied from other genes by DNA replicase. A protein called RNA transcriptase uses DNA as a template to create mRNAs, which then travel to the ribosome. The ribosome uses the mRNA as a template to create proteins with codons (three nucleotide segments) corresponding to individual amino acids. Transcription involves copying information from DNA to RNA; the language is roughly the same, so only transcription is required. Translation happens when mRNA codes for proteins; the language of nucleotides becomes the language of amino acids.

Although the central dogma may help us understand the process of information storage and application, its strictly dogmatic character has come into question. Originally, geneticists believed that all information flowed downstream: DNA→RNA→proteins. Within the last forty years, doctors came to understand that some viruses store their information as RNA. These viruses, called retroviruses, "reverse transcribe" the viral RNA, inserting new DNA into the host genome. The host then executes the instructions in the new DNA and manufactures new viruses. Retroviruses provided evidence that the central dogma theory is not really dogmatic.

LIFE IN SPACE

Information can be transferred back from RNA to DNA. So the central dogma does not always hold, but it does give us a general picture of how genes operate.

Lipids

Biochemists define lipids on the basis of solubility rather than structure. Lipids are insoluble in water but readily soluble in organic solvents such as ether and benzene. The importance of these properties might not be immediately appreciated, but they have a profound impact on life. Biochemistry, by and large, happens in aqueous solution. Oceans, lakes, and rivers, as well as the delicately balanced solution inside cells (cytosol), contain primarily water. Lipids form essential barriers between watery regions. They keep some chemicals apart and hold others together. When I discuss energy production in the next chapter, we will see how important membranes are to trapping and storing energy. Lipids can also be used for energy storage (fats and oils) and a number of other purposes.

Fatty Acids and Triglycerides

Many will recognize triglycerides as a regular part of daily life. These molecules make up common fats and oils.[16] A triglyceride has two components, a glycerol and three fatty acids (Figure 15.12). Glycerol forms a backbone out of a three-carbon chain; each carbon bears a single hydroxyl group capable of binding a carboxylic acid. Fatty acids form with a single carboxylic acid group ($-COOH$) at one end of a long, unadorned chain of carbon.

Fatty acids come in two flavors. Figure 15.13 shows both a saturated and an unsaturated fatty acid. A saturated fatty acid has as many hydrogens as the carbon chain can hold. In other words, all the carbon-carbon bonds are single bonds. Saturated acids tend to stretch out in almost straight lines, allowing them to pack neatly into crystals with high melting points; they are more likely to be solid at room temperature. In unsatu-

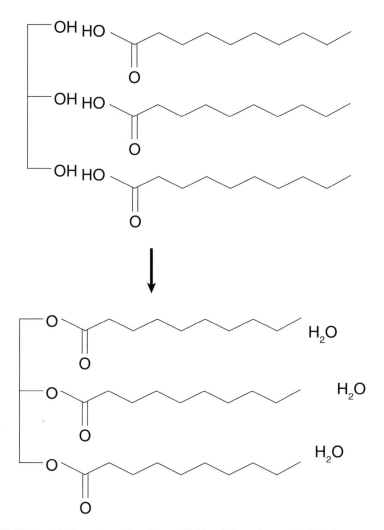

15.12 Triglyceride formation. One glycerol (left) and three fatty acids (right) go into a triglyceride. Triglycerides make up common fats and oils.

rated acids, two or more carbons are joined by double bonds. There are fewer hydrogens and a bend in the chain. The kink in unsaturated acids—from the double bond—makes them pack less neatly and, thus, have lower melting points. Unsaturated acids are more likely to be liquid at room temperature.

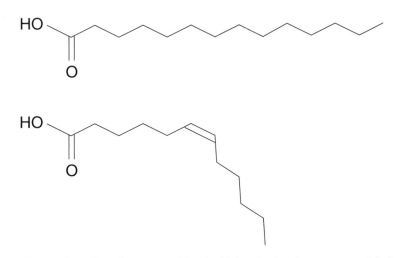

15.13 Fatty acids can be either saturated (no double bonds, above) or unsaturated (at least one double bond, below).

Organisms assemble fatty acids and triglycerides in order to store energy. Fatty acids hold energy more efficiently than carbohydrates. Oleic acid (with 18 carbons) can be broken up to yield 146 ATP equivalents, 28 percent more than three molecules of glucose (also 18 carbons). Plants and animals often store energy in this form. They pack energy-rich triglycerides into seeds and eggs to provide ready nutrients for the next generation. Nutritionists call solid triglyceride mixtures fats; they usually come from animals. Lard, butter, and beef tallow all have high percentages of saturated fatty acids. Liquid triglyceride mixtures are oils, more often associated with plants. Olive oil and other plant fats contain predominantly unsaturated fatty acids.

Phospholipids

A lipid with a phosphate group constitutes a phospholipid. If you were to take a triglyceride and replace one of the fatty acids with a phosphate group (Figure 15.14), this would produce a molecule known as a phosphoglyceride, the basis for most membrane proteins. A number of short-chain organic molecules can be attached on the other side of the phosphate

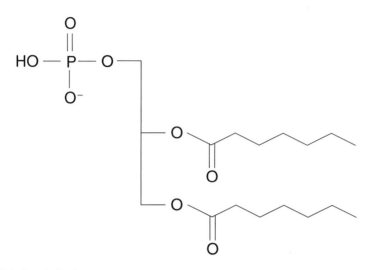

15.14 A phospholipid.

to add new functions, but the important properties of the molecules stay the same. One end of a phosphoglyceride is highly hydrophobic—the fatty acids repel water. One end is hydrophilic—the phosphate oxygens attract water.

Large numbers of phosphoglycerides in aqueous solution will begin to form bubbles. The hydrophobic tails stick together as they avoid the water, leaving small globules of lipid. As more and more lipids try to pack into the globule, the shape changes. More efficient packing leads to a double sheet (Figure 15.15). The top and bottom surfaces both have phosphate pointing out, creating a sandwich with the organic (carbon-based, hydrophobic) layer separating two aqueous (water-based, hydrophilic) layers. The ends of this double sheet curve around to form a larger sphere, with an aqueous center surrounded by a thin organic wall. Biochemists refer to the wall as a phospholipid bilayer or membrane.

Membranes serve a number of important functions. Cell biologists call the outermost membrane of a cell the cell membrane. It keeps the inte-

Aqueous (polar)

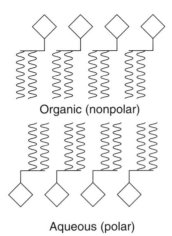

Organic (nonpolar)

Aqueous (polar)

15.15 A phospholipid bilayer.

rior separated from external conditions. Inside the cell, a number of other membranes form as well. In eukaryotic cells, membranes define the boundaries of a number of organelles, the nucleus, endoplasmic reticulum, and Golgi apparatus among them. Even in prokaryotic cells, internal membranes may be used to set up chemical reactions, separate reagents, and store materials. The membranes provide a fixed surface on which to set up proteins and through which to transport energy.

Waxes

Waxes occur when glycerol is replaced with ethylene diol, a two-carbon alcohol that binds only two fatty-acid chains. Such waxes include beeswax, spermaceti (from whales), and carnauba wax (from palm trees). Organic waxes also include a number of other molecules that combine long-chain fatty acids with long-chain alcohols. Waxes have many unsaturated fatty-acid chains, making them solid, malleable, and hydrophobic. For this reason, organisms often use waxes as a water-repellent coating on hair, feathers, and leaves.

Isoprene and Terpenoids

Several other categories of lipid have been named on the basis of small subunits that can be assembled into large molecules. These molecules, often containing double bonds and carbon ring structures, can be tremendously important because of the way all the carbons share electrons. The shared electrons give these lipids rigidity and allow them to participate in energy-sensitive reactions.

Isoprene, a simple five-carbon molecule, can be used to construct an amazing variety of structures, and Figure 15.16 shows several common variations. Two isoprenes joined together make a terpene. Terpenes' small size and shared electrons make them highly aromatic. Natural scents contain subtle mixtures of many molecules, but a terpene often dominates. Common examples include geraniol (roses), limonene (oranges), menthol (peppermint), and pinene (pine resin). Three terpenes can be assembled to make a steroid. Steroids appear throughout many organisms, but the two most famous are the human hormones estrogen and testosterone. Add a hydroxyl group, and you have a sterol such as cholesterol. Four terpenes (eight isoprenes) go into the construction of carotenoids, important molecules for photon capture in photosynthesis. Larger assemblages occur as well. Rubber is nothing more than long chains of terpenes strung end to end.

Lipids form something of a catch-all category. They do not follow a simple assembly pattern, as do the carbohydrates, polypeptides, and nucleic acids. Only a common fear of water holds them together. Lipids are mostly hydrophobic—though we have seen how the selective affinity for water of phospholipids makes them particularly useful. Fats, oils, waxes, steroids, and countless other molecules all fall into the lipid category.

In this chapter we have seen how the fundamental principles of life come into play in the biochemical details. Energy can be stored as reducing

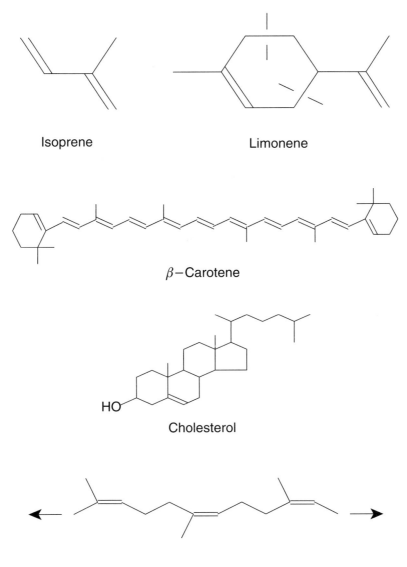

Isoprene

Limonene

β-Carotene

Cholesterol

HO

Rubber

15.16 Isoprene. This diagram shows isoprene and several terpenoids—molecules made from two-isoprene units. Limonene, a terpene, has one (two-isoprene) unit, whereas β-carotene has four units. Cholesterol has three, but has been modified by carbon removal and hydroxyl addition. Rubber forms as a long polymer of units.

power in the form of ATP (a nucleotide triphosphate), glucose (a mono-saccharide), starch (a polysaccharide), and fats (lipids). Life strings carbons together using rubisco (an enzyme), creating sugars that can be transformed into all the other organic molecules that make up life. Lipids form membranes that store water, keeping reagents together and creating an ideal environment for metabolism. We also have evidence of two principles discussed earlier. First, life obeys the second law of thermodynamics; more energy goes into metabolism (specifically glucose formation) than we can get out (glucose breakdown). Second, we have seen evidence for the hierarchical scale of life. A limited number of units are used in a variety of ways to generate a higher level of organization. Terrean life makes sense as an ordered, comprehensible whole.

16

Metabolism

We may not yet have a direct answer for how life on Earth began, but we have a good understanding of what keeps it going. Metabolism refers to the set of chemical processes occurring within living organisms, particularly those that string carbons together and break them apart. These processes, divided here into carbon chemistry and electron chemistry, either store or release energy, fueling life.[1] Carbon chemistry also can be divided neatly into two parts, anabolism and catabolism, putting life together and taking it apart. Anabolism involves stringing carbon together with other atoms to form complex organic molecules. It requires energy, a portion of which comes to be stored in the new molecules. Catabolism involves breaking organic molecules apart; it releases stored energy. Electron chemistry likewise can be divided into two parts, reduction and oxidation. Reduction involves the loss of electrons from a molecule. Oxidation involves the gain of electrons. One molecule must be oxidized for another to be reduced and, thus, the transfer of electrons can be called a redox reaction.[2]

Water will not appear prominently in the discussion that follows, but we must remember that water serves as a constant backdrop. It forms hydrogen bonds with carbohydrates, proteins, and nucleic acids. It makes them more likely to react with one another. Water repels lipids, causing them to form globules, sheets, and chambers. These structures anchor and separate aqueous solutions, making complex metabolism possible. Finally, many of the reactions we have already seen involve the loss or addition of water. Hydration and dehydration reactions occur regularly. Water provides the medium in which many metabolic reactions take place.

You Are What You Eat

All organisms can be categorized by their trophism—what they eat. We characterize metabolic reactions using words with the suffix "-trophy" (from the Greek, to nourish). These words describe how organisms get their carbons and electrons.

Autotrophs fix carbon; they take single carbon molecules—commonly carbon dioxide (CO_2) and methane (CH_4)—and incorporate them into larger organic molecules. Ecologists describe autotrophs as primary producers because they exist at the very bottom of the food chain. Their importance comes from all the species that rely on them for food. Humans, for instance, cannot incorporate single carbons. We must eat organic molecules created by other organisms to survive. Biologists call humans heterotrophs because we must eat other ("hetero-") organisms. Autotrophs make their own ("auto-") food. Heterotrophs eat autotrophs (or other heterotrophs). So the first distinction we find between organisms comes from their ability to "fix" carbon and make organic molecules. Autotrophs can; heterotrophs cannot.

Just as organisms take carbon, they also take electrons from organic or inorganic molecules. Organisms that take electrons from organic molecules are known as organotrophs; from inorganic molecules, lithotrophs. Lithotrophy would seem to imply that the electrons come from stone (*lithos*), but this is not always the case. Most lithotrophs get their electrons from sulfur and other reduced molecules in aqueous solution.[3] If an organism takes electrons, it must also dump them. Electron dumps regularly involve oxygen, either as part of an organic molecule or attached to another atom.[4]

LIFE IN SPACE

Redox potential measures the pull that a molecule has for an electron. Like any other form of potential energy, redox potential can be compared to a hill. The higher up the hill an object is, the more potential it has to roll downward, and the more energy can be released as it does. Electrons only loosely attached to their molecule can be easily transferred to another molecule. The first molecule is oxidized and the second molecule reduced. The first molecule, then, must have less pull—less affinity—for the electron in question. Chemists calculate redox potential by seeing how easily an electron can be pulled off of a molecule. Molecules with little affinity have low redox potential and can reduce any molecule with a higher potential. As the electron passes from molecule to molecule, it has less and less potential energy.

The only confusing part of this scheme comes from the fact that low redox potential corresponds to high potential energy. A molecule with low redox potential has a high-energy electron just waiting to reduce another molecule. An organism acquires a high-energy electron (from a molecule with low redox potential), uses the electron to reduce a number of organic molecules in sequence, and then discards the low-energy electron onto a molecule with a high redox potential. Redox reactions are driven by potential energy.

A final distinction in trophism relates to an organism's energy source. Chemotrophs use reduced molecules for energy. In other words, they have a chemical power source. Phototrophs use light energy. Phototrophs take low-energy electrons from their environment and excite them with photons. The pigment chlorophyll holds the low-energy electron in place while light energy charges the molecule.[5] The charged chlorophyll can then act as a very powerful reductant.

Are there sources of energy other than chemical energy or light? Astrobiologists sometimes speculate on whether life could use such energy sources as heat, pressure, or gravity. To date, however, all known organisms run on light or chemical energy. Summing up an organism's sources of energy, carbon, and electrons can be quite a mouthful. Cyanobacteria, for instance, are photolithoautotrophs. They get energy from sunlight, electrons and carbon from inorganic sources. Humans are chemoorgano-heterotrophs, getting energy, electrons, and carbon from organic molecules. The next few sections discuss specific metabolisms in more detail.

Phototrophy

Phototrophy has become the most important energy source for the terran biosphere, accounting for more than 99 percent of the energy budget. The term "photosynthesis" properly refers only to photoautotrophs, which run on light and synthesize organic molecules. Because the most common phototrophs (plants and cyanobacteria) are photosynthetic, you will often see the terms used interchangeably.

How does phototrophy work? Phototrophic organisms use sunlight to charge molecules, which can then reduce other molecules. Every phototroph depends upon large arrays of molecules to trap sunlight. A phototrophic cell contains areas devoted to catching photons. Carotenoids are one of the more common pigments used for this purpose. Their multiple double bonds mean that electrons can be shared among a number of different atoms and promoted to a variety of energy levels. This variability allows them to capture light at multiple wavelengths.

In terran life, three types of molecules absorb most of the light: carotenoids, phycobiliproteins, and porphyrins. All three use double bonds and ring structures to maximize their ability to capture photons. Carotenoids absorb light in the range of 400–600 nm. Phycobiliproteins, a class of proteins with a prosthetic group (a nonprotein component) called a phycobilin, absorb light in the 450–650 nm range. Porphyrin rings, including chlorophyll and bacteriochlorophyll, cover the 700–1050 nm range. Recall that shorter wavelengths have higher energy and that energy moves downhill. So carotenoids can pass their charge to porphyrins, but the charge cannot be passed back. The three types of molecules carry light energy to a central protein unit, the photosynthetic reaction center, much like a satellite dish transmits a signal to your television.

The photosynthetic reaction center consists of a pair of proteins. The proteins act as a scaffolding for six porphyrin molecules, which are always placed in the same relation to one another. The top two porphyrins are chlorophyll (or bacteriochlorophyll). When a photon hits a reaction-center chlorophyll (or when the energy arrives from the antenna), it excites a single electron and charges the molecule. The charged molecule immediately excites one of the other porphyrins, starting a cascade of redox reactions. In cyanobacteria, this chain ends with the reduction of a molecule called nicotinamide adenine dinucleotide phosphate. The name can be a bit daunting, so biochemists usually call it $NADP^+$. When charged, the molecule takes on an electron and a hydrogen molecule and becomes NADPH. (Despite the long name, we now have the ability to figure out something about the structure of this molecule. A dinucleotide, NADP has two nucleosides joined by phosphate groups, making it a modified nucleic acid.) In other phototrophs, the chain ends by reducing either nicotinamide adenine dinucleotide (NAD^+) or ferredoxin (a protein with associated iron molecules). All three molecules store energy for later use. NADH has the reducing power of 2.5–3 ATP and NADPH of 3.5–4 ATP. Ferredoxins vary, but some have the ability to reduce NADPH. Generally, one photon can be captured and converted into around 3 ATP.

In addition to storing energy in reduced molecules, some reaction centers pump hydrogen ions (H^+) across the membrane. I discussed the general principle of chemiosmotic potential as a way for cells to store energy in Chapter 6. As the ions build up on one side, osmotic pressure tries to push them back across the membrane. A special protein called ATP synthase allows the ions to cross but uses the energy of their passing to attach a third phosphate to ADP. Thus some phototrophs get extra ATP from the process. Later in this chapter I'll discuss how cells convert

NADH, NADPH, and ferredoxin to ATP through the electron transport chain, but first, let's take a look at photosynthesis.

Photosynthesis

Cyanobacteria and plants both fix carbon with the power generated by phototrophy. They use an anabolic pathway known as the Calvin-Benson cycle to incorporate the carbon from carbon dioxide into organic molecules. In truth, the Calvin-Benson cycle attaches carbon atoms to the ends of existing sugars, as this is easier than making whole new organic molecules. The full cycle involves fourteen kinds of sugar, ranging from three carbon chains to seven carbon chains, some of which can recombine in a number of ways. The long and short of it, however, is that six inorganic carbons (from carbon dioxide) turn into one glucose molecule at an expense of 54 ATP in energy. The Calvin-Benson cycle is not the most efficient method of fixing carbon, but it is by far the most common.[6]

Oxygenic Photosynthesis

The rise of oxygenic photosynthesis has affected Earth more than any other biological process. This one type of photosynthesis outstrips all others in efficiency and impact on the planet. Astrobiologists find it particularly important for three reasons. First, oxygenic phototrophs have had a bigger impact on the planet than any other group of organisms, including humans. Second, oxygenic phototrophs absorb light at shorter wavelengths than anoxygenic phototrophs, making them chemically powerful. Third, oxygenic photoautotrophs require nothing more than carbon dioxide, light, and water to live. This makes them more independent than any other form of life on Earth.

All of these reasons point toward one conclusion: cyanobacteria, the only oxygenic phototrophs, are the most successful group on the planet. Sometime around 2.5 billion years ago, an oxygenic revolution occurred and so much oxygen was produced that the atmosphere and oceans became oxidized. To our knowledge, no inorganic process could have this result. An oxidized atmosphere would be visible even from light years away with current technology. Astrobiologists have made understanding oxygenic phototrophy a priority. Not only did the production of oxygen result in changed atmospheric and oceanic chemistry, it dramatically affected the biosphere. Paleontologists believe that, before the oxygenic revolution, few organisms were adapted to live in the presence of abundant oxygen. Oxygen molecules acted as a poison, readily reacting with complex organic molecules and breaking them down. Most prior life would have been exterminated. At the same time, high levels of atmospheric oxygen resulted in a layer of ozone. Ozone stops very short wavelengths of light from hitting the surface, protecting the biosphere from their harmful effects. (Ultraviolet light has a particularly bad effect on nucleic acids.) Life outside the oceans may have been possible only because of the ozone layer.

The evolutionary innovation that made oxygen production possible seems minimal. Oxygenic phototrophs have two unique properties. I mentioned above that porphyrins absorb in the 700–1050 nm range and that shorter wavelengths were more powerful. Anoxygenic phototrophs—and presumably the oldest phototrophs—use bacteriochlorophyll in their reaction centers. These bacteriochlorophylls absorb light between 798 and 870 nm. The two cyanobacterial photosystems absorb light at 680 and 700 nm, giving them considerably more reducing power. A second important property comes from a protein complex called the manganese wheel. The wheel can strip four electrons from water, turning two water molecules into oxygen gas and hydrogen ($2 H_2O \rightarrow O_2 + 4 H^+ + 4e^-$). Water has a relatively high redox potential, making it a rare and useful source of low-energy electrons. Only cyanobacteria have evolved the ability to take and use the electrons in water. The two developments, powerful pigments and the manganese wheel, may or may not be related. They always occur together in living organisms, but this may not always have been the case.

This unique form of photolithoautotrophy—energy from light, electrons from water, carbon from carbon dioxide—meant that cyanobacteria could live anywhere. Previous organisms were limited by the presence of reduced compounds near volcanic outflows and deep-sea vents. They required easy or high-energy electrons. Cyanobacteria needed only sunlight, water, and carbon dioxide. Sunlight strikes everywhere on Earth and can still be strong enough to power photosynthesis up to 200 m below the surface of the sea.[7] Water fills the oceans, collects on land in rivers, lakes, and seas, and even permeates the atmosphere. Oxygenic photosynthesis meant that the biosphere could expand from a few microenvironments to cover the face of the planet.

Phototrophic Organisms

Phototrophic organisms come in three types, anoxygenic phototrophic bacteria, oxygenic phototrophic bacteria, and oxygenic eukaryotes. Surprisingly, groups of phototrophs do not seem to be closely related to one another, though they all use almost identical reaction center proteins. Four divisions of bacteria contain anoxygenic phototrophs. Divisions are kingdom- or phylum-sized groups of bacteria. The divisions Chloroflexi and Chlorobium contain only phototrophs.[8] Two divisions are predominantly nonphototrophic, but contain phototrophic groups. The Proteobacteria contain purple sulfur and purple nonsulfur bacteria.[9] The gram-positive bacteria (Firmicutes) contain a group called Heliobacteria. Phylogeneticists, who study the relations between organisms, remain baffled by exactly how these groups of phototrophs relate to one another. If you take a close look at Figure 17.3, you will see the position of all phototrophic bacteria on the tree of life.

Cyanobacteria, which utilize oxygenic photosynthesis, represent a large bacterial division with members present everywhere biologists have looked. Cyanobacteria dominate the ecosystem at the top of oceans and lakes. They form part of the symbiosis that makes up lichen, appearing on stone, concrete, and other hard surfaces. Cyanobacteria have been found floating in the air. They were located at the edge of a nuclear blast site. Cyanobacteria have even colonized the cells of other organisms, forming chloroplasts, the photosynthetic organelle in eukaryotes. Not just another group of phototrophs, phototrophic eukaryotes play host to cyanobacteria.

Chemotrophy

Phototrophy accounts for almost all of the energy and carbon entering the biosphere, but several other processes make interesting contributions. Chemolithoautotrophs live off of inorganic compounds, fixing carbon and bringing energy into the system. Far more common, chemoheterotrophs must live off of organic compounds generated by autotrophs.

Chemolithotrophy

Chemolithoautotrophy reflects the ability of organisms to live by getting carbon and charged electrons from inorganic sources. Many of them are anaerobic, tolerating only a little oxygen in their environment. This heavily restricts their ecological niche and means you are less likely to encounter them on a daily basis. Chemolithotrophs pick up electrons in a reduced state, taking energy and electrons from the same molecule. This molecule usually comes from some natural process of terran geology.

One of the most effective molecules is hydrogen gas (H_2), because of its low redox potential. Despite the incredible abundance of hydrogen in the universe, pure hydrogen can be hard to come by on Earth. Planetary scientists believe that hydrogen is so light (think of hydrogen-filled balloons) that most of the hydrogen gas in the atmosphere has been lost to space. It just drifted away. Hydrogen may be the third most abundant element at the surface of the planet, but it remains tied up in organic molecules and water.[10] Those organisms dependent upon hydrogen gas for fuel must rely on other organisms to produce it.

After hydrogen, sulfur may be the best atom for chemolithotrophs. A number of unicellular organisms make their living on reduced sulfur compounds. The most reduced version of sulfur comes from hydrogen sulfide (H_2S). Green and purple sulfur bacteria both use hydrogen sulfide, leading some microbiologists to believe that these phototrophs were precursors to the oxygenic photosystem. It seems a short step from oxidizing H_2S to oxidizing H_2O. Hydrogen sulfide also provides fuel for deep-sea vent ecosystems.[11] Most terran sulfur resides in rocks or sub-ocean sediments. Tectonic activity stirs up these materials, releasing reduced sulfur in the form of hydrogen sulfide. The molecule has a foul smell, responsible for the odor of rotten eggs, flatulence, and volcanoes. Brimstone, the medieval name for sulfur, gets its odor from hydrogen sulfide.

At hydrothermal vents, tectonic activity releases magma and gases at the sea floor. Sulfur-eating bacteria convert reduced sulfur into usable energy, fueling a whole ecosystem. These ecosystems do not extend far beyond their sources of energy, but they can be quite spectacular. Organic carbon, falling from the ocean surface, has been built into plants and animals unlike anything that grows in the sunlight. Astrobiologists suspect that the first ecosystems on Earth may have formed around deep-sea vents before the evolution of oxygenic photosynthesis. We also wonder whether deep-sea vents on icy moons like Europa might have all the necessary components for life.

As with hydrogen gas, hydrogen sulfide comes as a byproduct of life. Microorganisms living in animal intestines and brackish swamps take advantage of the available reduced sulfur. Several organisms have shown the ability to utilize elemental sulfur as well (S^0). Although this sulfur has a higher redox potential than hydrogen sulfide, it still has enough energy to run an organism.

A number of other reduced molecules can also be used, although none of them has as much energy as hydrogen and sulfur. Iron, nitrogen, and manganese have all contributed to the energy budget of life on Earth. Astrobiologists try to keep this in mind when looking for life elsewhere. Even though most of our energy comes from sunlight, other sources may be more important on other planets.

Chemoorganotrophy

Chemoorganoheterotrophy means that energy, electrons, and carbon atoms all come from organic molecules. One of the best arguments for a common ancestor to all living things comes from this type of metabolism. Almost every organism breaks down glucose ($C_6H_{12}O_6$) into two pyruvate molecules ($C_3H_3O_3$) in a way that produces 8 ATP equivalents worth of energy. The process seems to be a universal feature of life on Earth. Any number of organic reactions could be imagined that would release energy, but for some reason, this one process—called glycolysis—has become central. Even organisms that produce energy by phototrophy retain this pathway. It allows them to use stored energy at night or under stress. Energy generation by autotrophy (photo- or chemo-) should be viewed as an optional extra. The ability to generate energy from organic molecules is standard.

Pyruvate acts as a bottleneck in terran metabolism. Glucose and carbohydrates can be broken down into pyruvate, and pyruvate serves as a common intermediate in the anabolism (build-up) and catabolism (break-down) of amino acids. In short, when organisms need to get energy from organic molecules, they do so by breaking them down into short carbon molecules—pyruvate and acetyl-CoA, just one reaction away from pyruvate.

Pyruvate in turn can be degraded in a number of ways. Organelles called mitochondria within eukaryotic cells catabolize pyruvate using the tricarboxylic acid (TCA) cycle. The TCA cycle produces a small amount of ATP in addition to NADH. In the presence of oxygen, mitochondria can convert NADH into even more ATP using the electron transport chain. In the absence of oxygen, other processes like fermentation must take place.

The Tricarboxylic Acid Cycle

Eukaryotes, including all the multicellular organisms, process pyruvate using the TCA cycle.[12] First, they convert pyruvate to acetyl-CoA. This involves removing two oxygens and a carbon (to make CO_2) and replacing them with coenzyme A (CoA). Coenzyme A is adenosine diphosphate with an attached carbon chain that vaguely resembles a tripeptide. A coenzyme makes it easier for enzymes to recognize and manipulate the molecule. Next the coenzyme comes off and the two remaining carbons get attached to a four-carbon chain. Carbons get popped off one at a time until only four carbons remain, ready to take on a new acetyl group. Each time a carbon pops off, a molecule of NAD becomes charged. Several other reactions also produce energy so that, in total, the TCA cycle produces 15 ATP equivalents for every pyruvate. Including glycolysis from the last section, the breakdown of one glucose molecule into six CO_2 molecules charges 38 ATP.

The Electron Transport Chain

ATP acts as the power source for many enzymes as they facilitate chemical reactions. Earlier in this chapter I mentioned NADPH, NADH, and ferredoxin and told you that each one was the equivalent of a certain number of ATP. I never mentioned how that happens; a gap I now want to fill.

Cells generate ATP directly during glycolysis and a number of other processes, but more energy is released in the form of the reduced molecules. These include NADH (2.5–3 ATP), NADPH (3.5–4 ATP), and ferredoxin along with flavin adenine dinucleotide (FAD/FADH$_2$, ~2 ATP).[13] Cells convert these molecules to ATP through a process called the electron transport chain. The electron transport chain resembles a phototrophic reaction center without the light. In both systems, the passage of electrons down a long redox chain pumps hydrogen across a membrane. In photosynthesis, the final electron acceptor is a reduced molecule such as

NADPH. The electron transport chain starts with NADPH (or a similar molecule) and the final electron acceptor is oxygen. In both systems, hydrogen builds up on one side of a membrane. The hydrogen ion gradient forces ions through ATP synthase, turning ADP into ATP. The ATP is then free to move about the cell, charging enzymes and helping power the metabolism. A small number of reduced molecules can be used in other pathways, but for the most part, these molecules get processed to make ATP.

Fermentation

A funny thing happens in cells when the oxygen runs out. The cell continues to break down organic molecules into pyruvate to produce energy, but they can no longer run the electron transport chain. Organisms that usually use the TCA cycle and electron transport chain face a problem. Neither process runs without oxygen. NADH in particular begins to build up with no place to dump its electrons. Fermentation provides no new energy for the cell, but it gets rid of the extra electrons so that glycolysis can continue. Without converting NADH, glycolysis only produces 8 ATP per glucose (compared to 38 with TCA). Still, that means more ATP than the cell would have otherwise.

Two types of fermentation reduce pyruvate with NADH, lowering the concentration of both molecules and creating a space for more glycolysis. Alcoholic fermentation converts pyruvate to ethanol with the enzyme alcohol dehydrogenase and one NADH. Microorganisms that produce alcoholic beverages use this pathway. Lactic acid fermentation converts pyruvate to lactic acid with the enzyme lactate dehydrogenase and one NADH. The microorganisms that sour milk and produce yogurt use this pathway. Lactic acid can also accumulate in human muscles during anaerobic exercise. When you work out hard enough that your cells cannot get sufficient oxygen, pyruvate and NADH build up. The cells convert the pyruvate to lactic acid, which makes you feel sore. The soreness will last for a few days until the blood can carry the lactic acid away. It never ceases to amaze me that so few processes can explain such a broad range of biological events.

The Endosymbiotic Theory

Biologists like to say that eukaryotes have structural diversity whereas prokaryotes have metabolic diversity. Multicellular organisms are all eukaryotes; they have cells with nuclei. Multicellularity accounts for a huge range of structural diversity. No one would say that a whale and fruit fly look anything alike. More than that, though, eukaryotes have tremendous complexity in subcellular structure—cell walls and organelles and more. At the same time, eukaryotes are incredibly boring metabolically. With few exceptions, eukaryotes all have the same metabolism, and much of that is borrowed. Glycolysis and fermentation both occur in the cytoplasm, but phototrophy, the tricarboxylic acid cycle, and the electron transport chain only happen within special organelles.

Phototrophy in eukaryotes is always located within an organelle called a chloroplast.[14] Chloroplasts have an outer membrane that separates them from the rest of the cell. They also have a highly folded inner membrane that divides the interior into two sections, the lumen and the stroma. Photosynthesis pumps hydrogen ions into the lumen. ATP forms as the ions pass back into the stroma.

The TCA cycle and the electron transport chain always occur within organelles called mitochondria. Mitochondria also have an outer membrane and a highly folded inner membrane. Once again the inner membrane separates two regions, the matrix and the intermembrane space. Electron transport pushes hydrogen ions across the inner membrane into the matrix. ATP forms as the ions pass back into the intermembrane space.

For at least a hundred years, biologists have known that something strange was going on with these two organelles. Mitochondria and chloroplasts both have their own DNA. If lost, a cell cannot replace them. Their membranes and DNA resemble the cell membrane and nucleic acids of bacteria. In 1905, Konstantin Mereschkowsky (1855–1921), a Russian biologist, proposed that chloroplasts might really be bacteria that had been incorporated into eukaryotic cells. The idea, which came to be known as the endosymbiotic theory, was popularized by the American biologist Lynn Margulis in the 1960s, and the advent of molecular phylogenetics has shown the theory to be correct. Chloroplasts are more closely related to cyanobacteria than to their host cells. Mitochondria appear to be close cousins to proteobacteria, a diverse group that also includes purple phototrophs.

The proteobacteria provide an interesting rationale for the incorporation of mitochondria. All close relatives to mitochondria are intracellular parasites. They live by infecting living cells. Mitochondria apparently became trapped and lost their ability to live independently. If you want to peek ahead, Figures 17.2 and 17.3 show where the mitochondria came from and where they ended up.

The endosymbiotic theory means that eukaryotes are not even capable of carrying out respiration and photosynthesis without prokaryotic aid. Photosynthetic eukaryotes rely on internal cyanobacteria to process light into energy. The TCA cycle and electron transport chain—both of which dramatically increase the energy efficiency of eukaryotes—occur only within the membranes of internal proteobacteria. Eukaryotes rely on prokaryotes for their metabolic diversity. Free-living prokaryotes show a wide range of metabolisms, including photolithoautotrophy and chemolithoautotrophy, and use a variety of fuels. One species of bacteria even lives by removing chlorine atoms from chemical waste.[15] Another bacterium can switch between phototrophy and chemotrophy, lithotrophy and organotrophy as the need arises.[16]

Metabolism includes all the biochemical reactions in a living organism. This chapter has focused on how energy, electrons, and carbons find their way into the biosphere. Autotrophs fix carbon from one-carbon atoms (like CO_2) into organic molecules like glucose. Heterotrophs eat organic molecules produced by other organisms (often simply eating the organisms). Phototrophs use sunlight to produce ATP and store energy in reduced molecules like NADH. Chemotrophs use chemicals to the same ends. The electron transport chain converts the reduced molecules into ATP, which acts as the currency of energy exchange, powering reactions throughout the cell. This means that the biosphere gets almost all of its energy from cyanobacteria and other phototrophs, which exist almost everywhere, including inside other cells.

Although organisms abound beyond number, this same basic metabolic pattern applies to them all. The commonality strongly supports the idea that all organisms share a common ancestor; life on Earth represents a single example of life in space. This chapter covered the common metabolism for all life on Earth. The next chapter talks about how we characterize, categorize, and catalog its diversity.

17

The Tree of Life

No discussion of life on Earth would be complete without some attempt to look at the diversity and interrelation of organisms. Chapter 15 introduced taxonomy, the classification and naming of organisms. In this chapter, I will go into much greater detail, looking at the history of the field as well as the state of the art. We will look at several historical systems before diving into the most popular system today: molecular phylogenetics. We will explore the three domains—Archaea, Bacteria, and Eukarya—that constitute the terrean biosphere. The next chapter addresses a few exceptional entities that blur the lines between life and nonlife and, therefore, do not fall under the current classification scheme. Of course, names say something significant about the namer. They come deeply embedded in how we see the world and what we look for. Astrobiologists are keenly aware of how various forms of life are classified because often such classifications inform them about the origins, extent, and definition of life.

Biology textbooks often begin with taxonomy. Establishing a structure first allows discussions of evolution, structure, and biochemistry to occur in the context of many specific examples. Biologists tend to catalog diversity, dropping individual creatures in a series of bins, whereas astrobiologists are more concerned with the context in which life occurs. I have put off this chapter as long as possible in the hopes of presenting a picture of life as a planetary phenomenon. All terrean life operates in the same way, with the same bottom-up structure, using the same molecules and metabolism, and always interacting as a unit, a biosphere. Rare metabolisms, like hydrogen gas chemotrophy, exist only in the context of a larger community. Hydrogen gas usually happens only around other organisms. Likewise, the deep-sea vent communities depend upon fixed carbon sinking from the surface of the ocean. So take this as a sign that the end is near—the end of the book, at least. We can begin to see how the diversity and commonality of life plays out for particular organisms.

Animal, Vegetable, or Mineral?

At their core, all systems of classification must make some presumptions about what they classify and why. The question, "What is life?" informs each system in an important way. I imagine that the first groupings were organized around very simple questions. Is it good to eat? Does it move? Is it slow enough to catch? Will it eat me? Am I slow enough to catch? In this context, life was all about eating and being eaten. This may even still be true at a biochemical level, through the mechanisms of anabolism and catabolism. And yet, as humans became more sophisticated in our interactions with the environment, our classification systems also became more sophisticated.

The oldest recorded taxonomic system comes from Aristotle. It will probably be familiar to every reader in the form of a basic question: "Animal, vegetable, or mineral?" In Book II of *On the Soul,* Aristotle defines life. "It is self-nourishment, growth, and decay that we speak of as life." He goes on to make a comparison—soul is to body as sailor is to ship—and draws certain conclusions. Souls grant the properties of nutrition/growth/reproduction, perception, and intellect. Some organisms only have a nutritive soul; these are plants. Some organisms also have the power to perceive; they have an animal soul. Finally, some animals have intellect—humans. This divides the world into four categories: those with no soul (minerals), those with nutritive soul (plants), those with sensitive soul (animals), and those with rational soul (humans). For Aristotle, the sensitive soul has nutritive "potentialities." The rational soul has nutritive and sensitive potentialities. The definition of life—having a soul—clearly informs the nature of Aristotle's taxonomy.

Aristotle built his taxonomy around such distinctions. This had a profound impact on how he answered certain questions. When does life begin? At the quickening, when nutritive potentialities appear. When does human life begin? When the rational soul appears. The system makes reproduction the most basic property of life. The system also sets up a hierarchy of matter, with humans at the top and inanimate matter at the bottom. Some form of this hierarchy remains with us, even today.

In medieval Europe, the idea of hierarchical taxonomy developed into a more complex ladder of being that came to be known as the *scala naturae.* The ladder of nature mirrored the ladder of human society, with a few powerful, important individuals at the top and a mass of less-important individuals at the bottom. Scholars added God and angels to the highest level and divided lower levels to include fine class distinctions. The cosmos was viewed as a divinely ordered whole in the shape of a pyramid.

Thomas Aquinas further differentiated humans and animals. For Aquinas, the animal soul arises from the earth, but the human soul comes from God.[1] This distinction placed a notch in the ladder of creation that would endure in biology well into the nineteenth century and in popular culture up to the present day. Humans are defined as separate from animals; they possess a divinely inspired soul. I will touch upon this question once more in Chapter 19, but for the moment it must be noted that the heavy, dark line separating humans from other organisms came to be drawn in the Middle Ages. Aristotle presented a qualitative difference between sensitive souls and rational souls. By the time of Aquinas we see an ontological difference between the heavenly soul and the human body. Aquinas speaks of this difference in the context of Aristotle's three types of soul, but even this nod to the commonality of souls would disappear by the time of the Enlightenment.

Evolution

The roots of two major paradigms in western science can be found in the seventeenth century: natural theology and evolution. Natural theology pursues an understanding of God through observation of the natural world rather than by revelation. Roman Catholic and Anglican theologians affirmed that natural theology and revealed theology must agree if pursued rigorously. Protestant theologians tended to discount natural theology, distrusting human reason. For them, only revelation brought truth. Deists, on the other hand, moved almost entirely toward religion based on natural theology. Deism was popular in America, Great Britain, and France during the seventeenth and eighteenth centuries. Much of modern biology developed with the deist understanding of natural theology.

Efforts by European biologists to discover the variety and extent of nature were largely colored by the idea that knowledge of nature would lead to knowledge of God. European powers would often send biologists to distant parts of the planet to catalog and name different kinds of life. Societies were set up for understanding astronomy, geology, and biology.

The idea that species evolve and change—even the idea that different species shared a common ancestor—had been proposed in ancient Greece.[2] This concept of evolution could be roughly divided into two camps—those who believed in teleological evolution (change toward some end), and those who believed in mechanical evolution (change by a given mechanism). It is no wonder that natural theology and evolution influenced each other, defining a doctrine of progress. As the universe progressed, so people thought, things got better. Organisms became more complex and moved up the ladder. Among scientists and other intellectuals, this vision of progress colored the beliefs of those who followed Darwin as well as his opponents.

Nineteenth-century Europeans inherited two competing cosmologies: an eternally static universe and an evolutionary, improving universe. The two were constantly in tension, and their competition can be seen in the science of the day. Astronomers explored the notion of a steady state universe versus an expanding one; geologists considered fixed continents versus plate tectonics; and biologists debated special creation versus evolution. The eternal camp came to favor classification as the appreciation of a beautiful pattern, set down and permanently fixed. The progress camp came to favor classification that reflected the history of relationships—a family tree of life on Earth. It is important to note that other options exist, notably random change, but these two schools were at the forefront in Darwin's time.

Darwin's major contribution to the idea of evolution came in two areas. First, he proposed a mechanism for evolution in the form of natural selection. The full title of his book is *On the Origin of Species by Means of Natural Selection, or the Preservation of Favoured Races in the Struggle for Life*. Darwin showed that, in the presence of limited resources, a diverse population must compete for access. Inheritance of traits meant that the most successful competitors would pass on the traits that made them successful. Second, Darwin gave multiple examples from nature of how the idea might work. So, Darwin did not invent evolution (which he called descent with modification), he made it practical.

The debate between natural theology and revealed theology set the stage for a huge conflict over evolution. Those who rejected natural theology feared that a changing universe would be harder to follow. The rules laid down by revelation might be subject to change. Evolution became opposed to faith. The deists, on the other hand, felt attached to the idea of progress. Evolution was tied to a social program of improving species and a religious program of growth toward God. Once again, modern concepts of biology avoid either position, but we still see the effects of these arguments.

I have included this discussion of the history of classification because I think it says something interesting about astrobiology. The way we assemble the different fields—in this case philosophy and biology—has a profound impact on how we answer questions. The definition of life cannot be limited to just one field or the other. It impinges on both and relates to our understanding of humanity and the relation of humanity to other life. The study of life in space gives us the perspective to see how our answers to questions in different fields impact our overall understanding.

Prokaryotes and Eukaryotes

One scientific discovery that set the stage for modern taxonomy involves microbiology. The animal, vegetable, mineral divide favored by Aristotle was still popular in the eighteenth century. The founder of modern taxonomic nomenclature, Carl Linnaeus (1707–1778), was still dividing the world into these three categories. But his picture of the world would soon change.

Antonie von Leeuwenhoek (1632–1723) invented a powerful handheld microscope and used it to study one-celled organisms. He was the first person to explore and report on the habits of microorganisms in detail. Leeuwenhoek referred to them as "wee animalcules." Through the next three centuries, as more and more one-celled organisms came to the attention of biologists, they would find a larger and larger place in biological taxonomy.

Unicellular organisms were originally lumped together into a single category, or crudely divided into plants and animals, often based on color (it is green, so it must be a plant) or motion (it moves, so it must be an animal). Biologists began to assign taxa (kingdom, phylum, class, order, family, genus, and species) to groups of bacteria based on their metabolic properties, but relationships were still poorly understood and the taxa somewhat flexible.

Microbial nomenclature has undergone rapid change in the last two centuries. Christian Gottfried Ehrenberg (1795–1876) introduced the term "bacteria" in 1838. It comes from the Greek for "small staff" and originally described microscopic, rod-shaped organisms. A marine biologist named Edouard Chatton (1883–1947) was the first person to divide organisms into eukaryotes and prokaryotes. Eukaryotes described plants and animals, which have a membrane-bound nucleus and many internal structures. Prokaryotes described the bacteria and blue-green algae, photosynthetic microorganisms near the surface of the water.[3] The name displays the idea of evolutionary progress; "prokaryote" implies that these organisms existed before the development of nuclei. The group Bacteria was extended to include all prokaryotes.

In 1959, an American ecologist named Robert Whittaker (1920–1980) proposed a comprehensive taxonomic system that remains popular today. He took the three kingdoms of Aristotle (animal, vegetable, and mineral) and expanded them into a five-kingdom arrangement that recognized microorganisms. Three kingdoms were added to animals and plants and minerals dropped out. The system recognized that mushrooms and fungi are different from plants. It also created a space for unicellular eukaryotes (protists) and all of the prokaryotes (monerans). Thus, the five kingdoms are Animalia (multicellular eukaryotic heterotrophs), Plantae (multicellular eukaryotic phototrophs), Fungi (multicellular eukaryotic heterotrophs that absorb their nutrients—often living off decaying organic matter), Protista (one-celled eukaryotes), and Monera (prokaryotes). The five-kingdom model became the standard for many years. Until the advent of molecular phylogenetics, this was the dominant paradigm.[4]

Molecular Phylogenetics

In the wake of Darwinian evolution, biologists began to dream of a classification system that would represent true historical relationships between organisms. One group of biologists, the cladists, began to argue that all taxonomic groups, from species up to kingdom, should be defined by shared derived characteristics and common ancestry.[5] Every category would be limited to organisms that shared a common ancestor and had inherited a specific trait from that ancestor. Classification should follow evolution. An opposing camp argued that we cannot reconstruct all historical relationships perfectly. They believed that taxonomy can only reflect similarities between organisms and should not be based on theories about history. An innovation in tracing evolution led to the almost total victory of the cladists.

Modern taxonomy reflects evolutionary history; it does not, however, rely on shared derived characters. The field of cladistics was superseded by molecular phylogenetics. Molecular biologists discovered that they could reconstruct evolutionary relationships using nucleic acids and proteins. Every nucleic acid and polypeptide has its own sequence of monomers. If a number of organisms share the same gene, then their sequences can be compared to see who is related to whom. Close relatives should have more in common than distant relatives. Sequences vary between individuals, but the more closely related two individuals are, the more nearly identical their sequences should be.[6]

Molecular phylogenetics deals with the reconstruction of family trees (*phylo-*, trees and *gen-*, to make) on the basis of nucleic acid and polypeptide sequence data. As of a few decades ago, the Darwinian notion of common descent won out in taxonomy. The names and categories of organisms now reflect—or attempt to reflect—the place they occupy in the tree of life that connects all organisms. Taxonomy is history.

In the early 1980s, the work of the American microbiologist Carl Woese radically reshaped our understanding of life. Woese discovered a nucleotide sequence shared by all organisms. The small subunit of ribosomes is highly conserved—all organisms need to have a reliable translator, converting nucleotide sequences into polypeptides. Ribosomes show very little variation in their sequence, because they have to perform so efficiently. Innovation would be dangerous. This conservation means that phylogeneticists can still line up the gene sequence that codes for ribosomal RNA, even between organisms as diverse as humans and bacteria. The small subunit of RNA (SSU rRNA) proved to be a short, easily identified sequence that could be isolated from every organism. Woese and his colleagues began collecting SSU rRNA wherever they could find it.

The Three Domains

When all the sequences were compared, Woese and his colleagues discovered something amazing. The vast majority of sequence diversity was within the prokaryotes. Not only that, but the prokaryotes appeared to be divided into two groups, each with nearly the diversity of the eukaryotes. On this basis, Woese divided Monera into two super-kingdoms, Bacteria and Archaea, which he called domains. The other four kingdoms he consolidated into a third domain, Eukarya. This three-domain system currently holds sway among most biologists. Terran life can be organized according to evolution. Phylogeneticists construct a tree that shows the relationships between organisms. Then taxonomists assign names to branches on the tree. Major branches, such as the one connecting animals to all other organisms, receive large names, such as Kingdom Animalia. Smaller branches define smaller groups such as orders and species. Figure 17.1 shows the three-domain model of the tree of life. The three domains represent the most basic division among living organisms, with Archaea and Eukarya more closely related to one another than either is to Bacteria. Mitochondria and chloroplasts both crossed over from one lineage to another as the bacteria were incorporated into the cells of eukaryotes.

Phylogenetics is a historical science—it reconstructs past events—so no way exists to be absolutely sure that any given tree is correct. As biologists collect more and more data, the trees get better and better. The tree most commonly shown to connect all life on Earth comes from SSU rRNA, but other trees have been generated. Biologists have confidence that this tree is correct because it matches the data (even when you analyze it in different ways), because it resembles trees produced using other DNA and protein sequences, and because it agrees with observation.

Like any good science, the molecular phylogenetic scheme pointed out areas for new investigations. Within decades biologists discovered hundreds of new organisms by looking for SSU rRNA similar to those in the rarer bacteria and archaeons. Cell and molecular biologists also discovered physical characteristics that distinguish the domains—they have now been

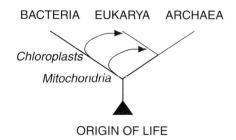

BACTERIA   EUKARYA   ARCHAEA

Chloroplasts

Mitochondria

ORIGIN OF LIFE

17.1 The three domains.

defended on the basis of shared derived characteristics as well as sequence data. The domains can be differentiated by RNA-polymerase (the enzyme that transcribes DNA to RNA), the starting amino acid in polypeptides, ribosome size and structure, and cell wall chemistry. In other words, the identity of each group extends all the way across organic chemistry: polypeptides, nucleic acids, lipids, and carbohydrates.

Eukarya

The most familiar domain is Eukarya (Figure 17.2).[7] It includes plants, animals, fungi, and protists. The first three groups have been slightly redefined to match up with evolutionary categories. The term protist, or occasionally protoctist, remains useful for referring to the remainder of Eukarya, but it does not reflect a clearly defined group. Protists, though often unicellular, can be multicellular, whereas plants and fungi, usually multicellular, can be unicellular. Eukaryotes all have internal membrane structure in their cells, including nuclei. Almost all have mitochondria as well. Eukaryotic cells carry their DNA in multiple, linear chromosomes and divide by processes called mitosis and meiosis.

Meiosis is the special type of cell division necessary for sexual reproduction. Some but not all eukaryotes are capable of creating gametes, cells with only half the genetic complement. Humans produce male gametes (sperm) and female gametes (eggs). Complementary gametes can fuse to form a new organism with a full genetic complement. Sexual reproduction

LIFE IN SPACE

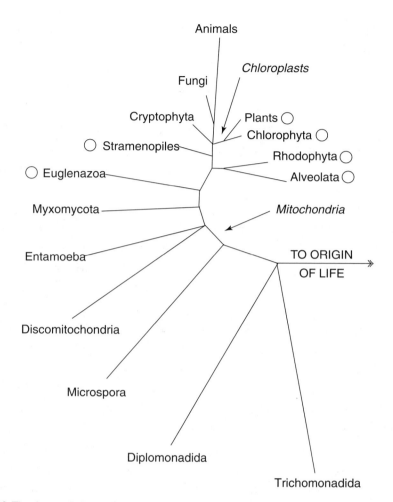

Animals

Chloroplasts

Fungi

Cryptophyta   Plants ◯

Chlorophyta ◯

◯ Stramenopiles

Rhodophyta ◯

◯ Euglenazoa

Alveolata ◯

Myxomycota

Mitochondria

Entamoeba

TO ORIGIN
OF LIFE

Discomitochondria

Microspora

Diplomonadida

Trichomonadida

17.2 The domain Eukarya. Eukarya contains the crown eukaryotes: plants, fungi, and animals (top). It also contains diverse phototrophs. Phototrophic groups are marked with a circle. Each one acquired this power by internalizing a bacterium or another eukaryote that had already internalized a bacterium. An arrow shows where a cyanobacterial relative was internalized to form the ancestor of all green algae and plants. Another arrow marks a proposed site for mitochondrial internalization.

allows for this random recombination of genetic material, increasing variation. Only eukaryotes have sex.

The most familiar eukaryotes fall into the kingdoms Animalia, Plantae, and Fungi. The three kingdoms still exist under the three-domain system. Informally, phylogeneticists refer to them as the "crown eukaryotes." The term recognizes that they form a very small, yet important set of branches on the tree. (It also represents a leftover element of the *scala naturae;* the crown is on top.) Surprisingly, fungi turn out to be more closely related to animals than to plants.

Eukarya, as mentioned in the previous chapter, have remarkably little metabolic diversity. For the most part, they are chemoorganoheterotrophs and feed off of the energy, electrons, and carbon already present in the biosphere. In other words, they eat other organisms. Phototrophy appears in a wide variety of eukaryotes, but always as a result of chloroplasts—trapped cyanobacteria. The number of times that this endosymbiosis (living together) evolved remains a little fuzzy. Chloroplasts—which all seem to be closely related—have been found in at least ten different phyla of eukaryotes—which do not seem to be closely related. Currently, biologists lean toward the idea that cyanobacteria were "domesticated" only once or twice. Other eukaryotes then acquired chloroplasts by taking in photosynthetic eukaryotes. Crudely, they ate the organisms that ate cyanobacteria. Evidence for this comes from the varying numbers of membranes. Chloroplasts have between one and five outer membranes, suggesting that some retain the remnants of a number of organisms. Almost all of the intermediate organisms have disappeared, but some nucleic-acid evidence remains, sandwiched between every other layer. Plantae and chlorophytes (green algae) form one group of phototrophic eukaryotes. Other groups include red algae and brown algae (both multicellular), dinoflagellates and euglena (both unicellular with flagella), diatoms (unicellular algae with glass shells), and kelp.

Mitochondria appear in all but a few eukaryotic species. Basal eukaryotes branch off near the base of the tree, meaning that they are the most distant relatives within the group. Basal eukaryotes lack mitochondria. This may mean that the mitochondria (close relatives of proteobacteria) were domesticated later, as suggested in Figure 17.2. It might also be the result of several early lines losing mitochondria. Phylogeneticists are divided on this.

A few additional protists (one-celled eukaryotes) deserve a word before we pass on to the other domains. Many groups, including amoebae and slime molds, have highly flexible outer membranes. Structural proteins inside the cytoplasm allow them to extend tentacles for movement and to engulf food. A few groups, including foraminiferans and diatoms, produce mineral shells for protection. The durability and complexity of these shells makes them excellent biomarkers. Many protists produce illness in humans, including beaver fever (*Giardia,* in water), sleeping sickness (*Trypanosoma,* carried by the tsetse fly), and malaria (*Plasmodium,* carried by mosquitoes).

Bacteria

The domain Bacteria contains most familiar prokaryotes, including cyanobacteria, proteobacteria, *Escherichia coli,* and the bacteria that sour wine (see Figure 17.3). Bacteria have no nucleus, but keep their DNA packed up in a region called the nucleolus. They utilize a wide variety of metabolisms. Five divisions (roughly the same as kingdoms) include phototrophs. A number of divisions also include chemoautotrophs, which derive energy from reduced molecules.

Bacteria have circular chromosomes and divide by binary fission. They duplicate internal structures, partition them between two halves, and split down the middle. Bacteria can rearrange their DNA by swapping pieces with other cells, but they cannot have sex. They swap DNA in a number of interesting ways.

Many bacteria possess an extra piece of DNA called a plasmid. A plasmid is like a mini-chromosome, but it contains nonessential information. It does nothing necessary for normal function in the cell, but it causes

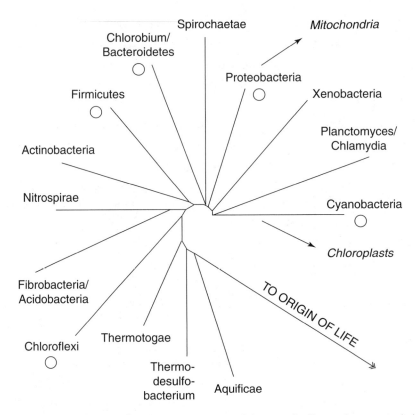

17.3 The domain Bacteria. Bacteria contains some of the most familiar prokaryotes including *E. coli*, *Y. pestis*, and cyanobacteria. It has five groups of phototrophs, each marked with a circle. Arrows show the origins of chloroplasts (close to cyanobacteria) and mitochondria (within the proteobacteria).

the host to copy it. It may also confer certain advantages on its host. Antibiotic-resistant bacteria, for instance, often get their resistance from genes on a plasmid. Plasmids may also have genes that code for conjugation, the ability to build a bridge to another cell. The bridge acts as a conduit so that a copy of the plasmid can be delivered to the second bacterium. This seeming self-interest leads some evolutionary biologists to question the meaning of individuality in biology. A plasmid includes DNA acting in its own best interest and not in the interests of the organism.

Bacteria can also transfer DNA by means of viruses and viruslike capsules. A virus might incorporate a small amount of host DNA when it takes over a cell. That information can be carried on to other cells in later infections. Stranger still, some proteobacteria appear to have domesticated virus genes. These genes form viruslike capsids and include randomly selected genes to be inserted and carried to other bacteria.[8]

Different systems have been proposed for the classification of bacteria. Major groups usually go by the title division or phylum. Current indecision about taxonomy, reflecting the rapid expansion of knowledge, means that groups with similar diversity to the Kingdom Fungi may still have the name of an order or a family. Roughly fourteen divisions are now recognized.

Although the individual species of bacteria may not be familiar, they play an important role in our daily lives. Bacteria live almost everywhere you can imagine and contribute their metabolic facilities to numerous human endeavors. Bacteria reside in all animal digestive systems (along with many protists), where they help digest food. Bacteria modify the flavor and chemistry of wine, chocolate, and coffee.[9] Cheese, sour cream, and yogurt all result from bacterial digestion of milk and cream. Most bacteria living in and on the human body are beneficial or harmless, but a few cause disease. Bacteria-related illnesses range from pneumonia (often *Streptococcus pneumoniae*) to tuberculosis (TB, *Mycobacterium tuberculosis*) and bubonic plague (the Black Death, *Yersinia pestis*). International treaties limit weapons research, but at times bacteria have been bred as weapons. Anthrax *(Bacillus anthracis)* made the news in the United States recently as a possible biological weapon.[10]

Archaea

Archaeons were practically unknown before Woese's work (Figure 17.4). Microbiologists thought of them as odd members of the bacteria. The few identified species produced methane gas or only lived at extremely high temperatures. Like bacteria, they have circular chromosomes and divide by binary fission. At the same time, archaeons pack and unpack DNA in some of the same ways as eukaryotes. Archaeons almost universally live in

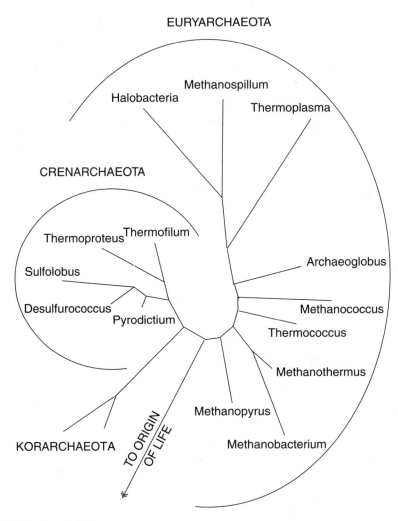

EURYARCHAEOTA

Methanospillum

Halobacteria

Thermoplasma

CRENARCHAEOTA

Thermoproteus Thermofilum

Archaeoglobus

Sulfolobus

Desulfurococcus

Methanococcus

Pyrodictium

Thermococcus

Methanothermus

KORARCHAEOTA

TO ORIGIN OF LIFE

Methanopyrus

Methanobacterium

17.4 The domain Archaea.

environments we would consider inhospitable. In astrobiology it can be easy to confuse Archean (a geological eon) with Archaea (a biological domain).[11]

Phylogeneticists divide Archaea into three branches. Roughly half of the species fall into each of two large groups, the Euryarchaeota and the Crenarchaeota. Many of the Euryarchaeota are methanogens; they produce methane as a byproduct of their metabolism. Methanogens quickly die in the presence of oxygen and may be a remnant of pre-oxygenic revolution Earth. Some are chemolithoautotrophs, using hydrogen gas as a source of charged electrons and carbon dioxide as a source of carbon. Euryarchaeota also contains a number of halophiles, organisms thriving in high-salt environments.

A few species within Euryarchaeota possess a proton pump similar to the light receptors found in the human eye, but no archaeons have been seen with phototrophic reaction centers like those in bacteria. The protein bacteriorhodopsin mediates the conversion of light into chemical energy (compare rhodopsin in the human eye). Microbiologists do not know how much of the archaeon's energy budget comes from sunlight. This makes it difficult to say whether they deserve the name phototroph. Recent discoveries show genes coding for bacteriorhodopsin also appear in some Fungi and Bacteria species, though the expression and function of these genes remains unknown.

The other major category, Crenarchaeota, live in extremes of temperature. Most cultured species thrive at high temperatures. Ecologists call these "heat-loving" species thermophiles.[12] The hottest on record have been observed to reproduce at 121°C (250°F). Thermophiles often have sulfur-based metabolisms (chemolithotrophs) and live at deep-sea vents and hot springs where reduced sulfur abounds. The high pressure at these locations keeps H_2O liquid even at extreme temperatures. Most organisms start to break down above 60°C (140°F) because their proteins lose their shape as bonds break. Thermophiles have developed special proteins that manage other proteins and DNA, keeping them functional.

Crenarchaeota also contains psychrophiles, cold-loving organisms. Some psychrophiles live in tiny pockets of liquid water within sea ice. The buildup of molecules in their cytosol acts as antifreeze and keeps the cells alive. Microbiologists are still exploring the most basic areas of psychrophile biochemistry and taxonomy, but organisms have been reported living at -20°C (-4°F).[13]

A very small number of organisms have been placed in the Korarchaeota, a branch at the very base of the Archaea. These organisms have not been cultured. Biologists only know of them because of SSU rRNA genes found in various locations. Other genes have been identified as well, but for the most part, the Korarchaeota remain a mystery.

Some astrobiologists wonder whether Archaeons might represent remnants of the first life on Earth. Their intermediate structure between Bacteria and Eukarya, their dislike of oxygen, their love for extreme heat, and their chemolithoautotrophic abilities all point toward early Earth environments. I am not sure whether I find this line of reasoning convincing, but it makes interesting food for thought. What were the earliest organisms like?

Rooting the Tree

Phylogeneticists would really like to determine a definitive root for the tree of life. Current evidence places it somewhere on the branch joining Bacteria to the other two domains (see Figure 17.1). The precise location, however, remains elusive. Each of the three domains connects to the others by a long branch, and long, unbroken branches always trouble phylogeneticists. They make the tree-producing algorithms behave a little strangely. A few studies have looked at genes that probably became duplicated prior to the last common ancestor of life on Earth. Using these genes, we can theorize a root such that the first division occurred between Bacteria and everything else. Archaea and Eukarya would have divided after that. These events may have been lost to time, though, and only minimal evidence has been found.

In this chapter we looked at the theory and practice of classifying organisms. I discussed just a few of the philosophical dilemmas that inform and bias our decisions with regard to defining and classifying life. It appears that we cannot escape our philosophical predilections. Still, it pays to be aware of where we started and what questions led to the answers we now have.

What biases appear in the current system? That is a very good question. The most fundamental assumptions are often the hardest to see. After all, we do not think of them as assumptions, but as basic facts. Astrobiology attempts to explore the assumptions of various fields by putting them in dialogue with other fields.

18

Exceptions

We have now come full circle. In Chapter 4 I asked how we define life. Although many interesting possibilities have been proposed, we saw that the basic answer is this—life is something like what we see on Earth. The most common definition, which I have called the pornography definition, said: "I can't define it, but I know it when I see it." This chapter explores whether or not that really is the case. Here we turn to specific examples from biology: viruses, prions, plasmids, and selfish genetic elements. Each entity has some but not all of the features we are looking for. Throughout this chapter, keep in mind these three questions: Whatever it is, is it alive? What does my answer say about my own definition of life? What does this entity tell me about life on Earth?

Viruses: Encapsulated Nucleic Acids

Viruses consist of nothing more than encapsulated nucleic acids. They carry their genetic component on one or a few molecules of either DNA or RNA. Proteins form a capsid or shell that protects the nucleic acids and helps to introduce them into cells. Some viruses have an additional envelope of lipids. Viruses prey on living cells, tricking them into accepting the viral nucleic acids and expressing their genes. The viral genes co-opt the host's metabolism to make more and more copies of the virus. A virus may make these copies immediately or lie dormant for a period of time before replicating and destroying the host cell. In the case of viral envelopes, the new viruses appropriate a small portion of the host's membranes.

Most viruses range from 10–300 nm in size, making the largest virus the size of the smallest cell. Most biologists do not consider viruses to be living, but active debate continues. In the argument for life, viruses undergo evolution and change, they utilize resources, and they can survive independently. In the argument against life, viruses have no cells and require a host to reproduce. So they meet the thermodynamic, evolutionary, and biochemical definitions of life, but not the structural or behavioral definitions.

Viruses cause a wide variety of human diseases, from colds (adenovirus) to AIDS or acquired immunodeficiency syndrome (human immunodeficiency virus, HIV). The latter is a great example of a retrovirus. HIV and other retroviruses use an enzyme called reverse transcriptase to create DNA based on an RNA template. These viruses generated a lot of interest because they violate the strict version of the central dogma theory.

Three other quasi-organisms resemble viruses in their general structure and way of life. Viroids exist as RNA without a protein coat. A single-stranded circular chromosome stores information and may catalyze reactions. Viroids were first discovered a few years ago as a disease agent in plants. In brief, they are viruses without capsids, or protective shells. Virusoids or virus satellites occur as short stretches of RNA that take advantage of both a host cell and a virus to replicate themselves. Like viroids, they exist as nothing more than opportunistic genes. Unlike viroids, they require the assistance of a virus to reproduce. Gene transfer agents (GTA) act as domesticated viruses in proteobacteria. Individual bacteria randomly incorporate DNA, encapsulate it, and send it out to be picked up by other bacteria. Like viruses, viroids, virusoids, and GTAs blur the lines between species, between individual and environment, and between life and nonlife. Nucleic acids alone can be evolutionary units.

Prions: Killer Peptides

Like nucleic acids, lone polypeptide chains have been known to cause disease. All known prions form with an irregular conformation of a normal protein. When a prion enters the environment of a cell, it acts to change the shape of common membrane proteins. The changed proteins string together to form long filaments within the brains of mammals. These long filaments cause damage by breaking through cells and tissues. Proteins with prionlike properties have also been found in fungi, though their function and effect remain unknown.

Prion-based diseases all have the same result—holes throughout the brain caused by the filaments. Because of this, pathologists (scientists who study disease) call them spongiform encephalopathies (literally, sponge-shaped brain diseases). Famous examples include Creutzfeldt-Jakob disease (in humans), scrapie (in sheep), and mad cow disease (in cows). Currently these diseases cannot be cured and result in rapid death. Methods of transmission are poorly understood, but some progress has been made toward a vaccine.

Plasmids: Acting in Their Own Best Interest

Some bacteria contain small stretches of DNA that act as extra chromosomes. The plasmids contain no essential genes, but may confer advantages to their host cell. The most well-known advantage comes in the form of antibiotic resistance. A number of plasmids code for genes that make their host immune to antibiotics such as penicillin. Doctors have become increasingly concerned about a biochemical arms race in which humans develop new medicines to kill bacteria and bacteria evolve new ways to resist. Plasmids represent the best bacterial weapon in the conflict.

Plasmids may be linear or circular and occur in eukaryotes and bacteria. Most are double-stranded, circular DNA. They contain a series of nucleotides called an origin of replication that acts as a starting point when the plasmid is copied by host proteins. Other genes on the plasmid often confer some benefit to the host. Antibiotic resistance has been mentioned. Some plasmids allow a bacteria to conjugate—form a bridge to another bacterium and send genetic material across. Some code for toxins that can be used to poison other bacteria. Some allow the host to break down unusual molecules like toluene that would otherwise be indigestible. Some plasmids exist only in limited copies within their host, while others replicate extensively. Clearly, wide variety exists.

My favorite property of plasmids comes from their ability to form addiction systems. Several known plasmids code for a long-term poison and a short-term antidote. As soon as a cell becomes infected, it starts expressing plasmid genes. The poison builds up within the cell, but is counteracted by an antidote. Both molecules come from the same source. The antidote, however, degrades rapidly, making the host dependent. If it loses the plasmid, or manages to stop following the genetic instructions, the host will run out of antidote before it runs out of poison and then die. Plasmid addiction systems make a strong case for gene-based selection. Evolutionary forces can act to preserve the information in a nucleotide sequence at the expense of cells and organisms. Once again we see that the notions of individual, organism, and species can be problematic.

Plasmids are not the only form of selfish genetic element. Geneticists have discovered that a number of nucleotide sequences within cells can force their own duplication. This can take the form of extra-chromosomal molecules like plasmids. It can also take the form of stretches within chromosomes that move about, become duplicated, and fill up available space.

The Origin of Life

All of these pseudo-organisms provide an important service to astrobiologists. They allow us to look at life not-as-we-know-it. Life seems to be such a complicated phenomenon and, as we question the transition from nonlife to life, we need to think about possible intermediaries. We need to see if life could operate somehow on a more modest scale.

Ribosomal catalysts, such as those found in ribosomes and viroids, have given astrobiologists hope that we may someday understand the origins of life. One of the greatest difficulties has come from our inability to imagine an organism that did not have two of the four molecular components of life. Nucleic acids were necessary for information storage and proteins were necessary for doing work. Ribosomal catalysts, called ribozymes, can do both. Many astrobiologists believe that the first life on Earth came in the form of a ribozyme that both carried information and replicated itself. Such a nucleic acid has yet to be developed, though in theory it remains possible.

Early in this book, I expressed my own doubts about the origin of life. I feel strongly that science can tell us important things about when and where this event happened. At the same time, our current knowledge remains extremely limited. A number of hypotheses have been put forward.[1] The most popular, expressed above, has been called the RNA world hypothesis. Proponents say that at one time the biosphere consisted only of ribozymes. These ribozymes existed in a membrane-free, open biochemical soup. They replicated themselves, growing more and more complex, and eventually developed into cellular life as we know it. Another hypothesis, called the protein world, postulates the presence of polypeptides acting as catalysts and information-storage molecules. Something like a prion may have been the first life on Earth. Finally, a few biochemists advocate for a lipid world. In the lipid world, membranes developed before other biochemistries, dividing by abiological processes in a way that led to evolution. This theory has the least adherents, but it would explain how life could generate the intense molecular concentrations that seem to be necessary.

Biochemists have demonstrated the formation of complex organic molecules. Most famously, the Miller-Urey experiment has shown that abiological processes can form long carbon-chain molecules such as amino acids. Other organic molecules appear in meteorites, the result of long-term chemical processes in space. Still, biochemists cannot account for the high concentrations necessary for living systems. Lipid bubbles, ion attraction, and evaporating pools all present possibilities, but none show definitive promise. Even should the right concentrations be possible, the molecules would all have different handedness. What (other than life) could produce the all-D sugars and all-L amino acids we see in the biosphere? We simply do not know how these molecules could have become organized in the complex pattern of modern life. The leap from nonlife to life, no matter how defined, seems extremely difficult.

The idea of bottom-up organization gives us hope that this problem may one day be solved. We know that a system like Earth, with huge quantities of energy pouring in daily, will be agitated in ways that result in more and more complex systems. I would never argue that the problem cannot be solved, simply that data are limited, so it has not. Research needs to continue precisely because we know so little and the question is so important.

The Definition of Life

These issues should cause us to reexamine our ideas about the definition of life. Most people, including most scientists I have known, view life as an all-or-nothing proposition. Either an organism is alive or it is not. Either a planet has life or it does not. It all makes sense intuitively. Just like a fire, an organism just barely alive can be brought back. So the life/postlife (life/death) distinction has been very successful for us. We get into trouble, though, when we try to extend this clarity into a prelife/life distinction. Debates about abortion revolve around this question. When does a line of cells within the mother take on the full status of human person? Indeed, theologians and philosophers have been debating the question for at least three thousand years. Somehow it does not come out as cleanly as life and death.

Sex makes the issue rather complicated. Bacteria and archaeons divide by splitting in half. Some eukaryotes, on the other hand, reproduce by way of sex. In sex a diploid organism, such as a human adult, has twofold nucleic acids (humans have forty-six chromosomes). The organism produces haploid gametes (sperm or eggs) with only half the nucleic acids (human gametes have twenty-three chromosomes). One gamete of each kind fuses to form a new diploid organism.

Biologists almost always describe it that way. Savvy readers will have noticed something odd about the description. Why is it that the diploid warrants being called an organism, but the haploid is called a gamete? Even biologists tend to view sperm and eggs as something less than individuals, something less than organisms. They have enough information to encode a new organism. They have energy, electrons, and carbon. They include carbohydrates, proteins, nucleic acids, and lipids. They even come in the forms of cells with cell membranes.[2] What else could be expected? So even biologists can fall into the trap of not thinking through the definition of life and of individuals.

The nineteenth-century novelist Samuel Butler (1835–1902) put it quite succinctly: "A hen is only an egg's way of making another egg." I think this says something important about the way we view life. Somewhere between the anthropic principle and the Goldilocks principle, we find a principle of familiarity. We acknowledge the hen and not the egg because we ourselves resemble the hen. When we think of life, we think of diploid hens, not haploid eggs. We think of mammals, not bacteria. We think of cellular organisms, not selfish bits of RNA. This colors our definitions of life.

Back to the egg. We tend to think of life starting in humans when gametes fuse to make a new diploid individual.[3] Why is this? The gametes seem to meet all of our definitions quite well. They even look like one-celled eukaryotes. So the prelife/life distinction has not been as well defined as the life/postlife distinction. We can even extend this to the origins of life on Earth. If life and nonlife are watertight compartments, then it should be impossible to get from one to the other. We will need to construct a hatch—an area where things (for want of a better term) can exist in between the two states. Our definition of life will need to be something other than "not dead." Viruses, prions, and selfish genetic elements exist in the murky area between life and nonlife. They have some but not all of the properties we want.

Since we are nearing the end of the book, I would like to introduce two of my own speculations about life. They do not have the same weight as earlier observations. They reflect only my own pondering of a very difficult question. I hope that my speculations will inspire you to ask questions of your own.

My first thought follows medicine. Doctors will often identify a syndrome when they do not know the cause of an illness (AIDS, for instance, was coined before HIV was discovered). A syndrome describes a collection of symptoms. It could be that we cannot yet identify a *sine qua non* (essential character) for life. We do not know what causes it or what it means. We can only identify a number of symptoms of life—things like evolution, biochemistry, and complexity. It may be possible to make a list and say that something is alive if it has three out of five of the symptoms. Or nine out of ten. The definition need not be so clear cut as the definition of a proton.[4]

My other thought arose in the context of statistics. It may be possible to define life as a process rather than a property of an individual. We have come to think of individuals and planets as having the property "alive" or the property "dead." Some definitive unit—be it cell, organism, or species—could be removed from the whole and subjected to testing. This would reveal whether or not it was alive. The process on another planet may not be so straightforward. What would happen if we chose the wrong unit? What if we chose the wrong test? A process-based description would look at life as a chain of variation. Energy, trapped in the system in the form of biochemical processes (metabolism) explores a variety of states as time passes. Natural selection ensures that more and more energy becomes stored within the system. A useful process-based definition would require a more sophisticated mathematical model than presently exists.[5] It could turn out to be useless practically, but it would force us to define organisms on the basis of their relationships to ancestors and offspring rather than on their relationship to their environment.

19

Intelligence

What book on life in space would be complete without at least some discussion of intelligence? It may be the most important property of life on Earth. I doubt that humans have altered the planet's chemistry as dramatically as the cyanobacteria, but we have accomplished two things that make us stand out. First, we have flooded space with information. Humans generate and broadcast tremendous quantities of information into space by way of radio and television and other signals. Second, humans have sent robots around the solar system and into deep space. We have landed instruments on Venus and Mars. In 2006, the *Voyager 1* spacecraft reached a distance of 100 AU from the Sun (roughly 15 billion km). This put it beyond the farthest planets, beyond the Kuiper belt, reaching into the Oort cloud. And it continues moving outward. Something about humans is different. From an evolutionary standpoint, the ability to reach other planets marks a major step.

Separating humans from other species seems like an intuitive process to us. Humans have created a number of categories to describe the difference. I named the chapter "intelligence" because that is the term we most often associate with extraterrestrials. The idea of extraterrestrial intelligence has long fascinated explorers and authors. Other names, however, have been associated with the magical property of humans that seems to separate us from all other species. Sapience and sentience have been used, and Earth's most popular religions all posit some notion of human uniqueness. Christians and Jews speak of humans being created in the image of God (Genesis 1:27). Muslims consider humans to be God's vicegerents, or deputies, on Earth (Sura 2:30). Hindus and Buddhists perceive the human experience as a unique occasion for enlightenment. We like to think of ourselves as special.

As an evolutionary biologist and phylogeneticist, I am inclined to think of humans as only a small branch on a very large tree—not particularly different. Up until this chapter, I have not introduced any criteria that would allow us to separate humans. Humans require water, carbon, and energy. Humans have eukaryotic cells and operate structurally, biochemically, and reproductively like other mammals. Humans are chemoorgano-heterotrophs, getting energy, electrons, and carbon from organic molecules produced by other organisms. With the exception of our ability to extend our influence into outer space, there does not seem much to set us apart.

Carl Linnaeus, the great taxonomist, listed humans as *Homo sapiens,* the wise man or the man able to apply knowledge. This chapter explores some very basic concepts related to intelligence and human distinctiveness. I recognize that humans represent something special, but I am still a little fuzzy on just what that might be. As with life, we find that making a clear and constant definition can be troublesome.

Astrobiologists rarely incorporate data from psychology, anthropology, or sociology, the fields that have traditionally studied and defined intelligence.[1] Nonetheless, I feel that no portrait of life on Earth would be complete without looking at the peculiar character and impact of intelligence. I also think that the idea of human uniqueness plays an important part in modern culture. Bearing this in mind, I would like to comment on what the ability to define intelligence might mean to the study of astrobiology.

What is intelligence? We use the word all the time, but what exactly does it mean? The *Oxford English Dictionary* defines intelligence as "the ability to acquire and apply knowledge and skills."[2] Such a broad definition would surely include the ability of chimpanzees and apes to learn sign language as well as the ability of bees to transmit information about the location of flowers. Perhaps intelligence, in and of itself, does not meet our requirements for distinguishing humans.

This does not mean that we should give up on the term. Beyond its basic definition, we use "intelligence" to describe a wide variety of phenomena. It has been connected to the ability to reason, although I am not sure that helps us much. Reason itself is often called into question when we look too closely at what we consider reasonable or unreasonable. One person's logic may be another's folly. We need to be more specific. Intelligence has been associated with the ability to solve problems and plan ahead. This capacity has been demonstrated by a number of mammals and birds. It might also be associated with the ability to think abstractly or to use language. Both of these traits have been more debatable. What constitutes abstract thought? What constitutes language? It seems to me that we could always adjust these definitions to ensure that humans were the only members of the group. But then that severely limits the term's usefulness. What we really want is something that would force us to say, "Yes, that thing is intelligent," even when we wanted to believe otherwise.

The sociologist Max Weber said it best: "The primary task of a useful teacher is to teach his students to recognize 'inconvenient' facts."[3] The most important results of intelligence, reason, and science come from their capacity to make us see things we don't want to see. Astrobiologists need to have definitions that allow us to be surprised and discomforted by our discoveries—definitions that make us ask the right questions. On the one hand, we need to calibrate our definitions of intelligence (and life) based on what we know. On the other hand, we need to allow the definitions enough latitude to show us what we have not yet seen.

The next few sections look at a few of the more common definitions of intelligence and human singularity. How do they inform your own definition of intelligence, and how do they relate to the definition of life?

Sentience

Some authors like to refer to humanlike species as sentient. The word sentient is a close relative of the verb "to sense" and has taken on several meanings related to sensation. In one interpretation sentience applies to all organisms that have physical sensations like sight, hearing, touch, taste, and smell. Another interpretation leans on sensation or consciousness as feeling. Sentient organisms feel emotions. The subjective nature of emotional feeling means we cannot be sure to what extent it matches up with physical feeling. The latter can be measured. The former cannot.

Physical Sentience

Astrobiologists benefit from discussions of physical sentience because they help to reveal our beliefs about the nature of life. Following the Goldilocks principle, humans tend to highlight familiar senses—touch, for example—and discount unfamiliar senses—such as the ability of birds and bacteria to detect Earth's magnetic field. When we consider life elsewhere, we need to bear in mind that forms of life, forms of intelligence, and forms of communication may depend upon different senses. An awareness of sensory variety on Earth should be a good preparation for detecting sentience elsewhere.

Humans' most trusted sense is sight. Vision happens much the same way as photosynthesis. Eyes use vast arrays of light-absorbing proteins. The proteins sit within membranes and pass signals to the brain. Animals use an enzyme called rhodopsin, consisting of a single polypeptide (opsin) and a two-terpene lipid (retinal). When a photon strikes the lipid, it changes conformation and starts a chain of reactions that closes sodium channels in a nerve cell. As a result, an ion gradient (Na^+) develops that sends a message to other nerve cells. Simple building blocks (polypeptides and lipids) and simple processes (photon chemistry and ion gradients) are used to build both photosynthesis and vision.

Humans have rods and cones, two types of nerve cells using rhodopsin.

Rods sense light brightness and absorb light around 510 nm in wavelength. Rod cells have the ability to amplify a signal so much that a single photon can close a million sodium channels. Rods allow us to see in dim light, but they are linked to one another and give poor resolution. Cones have slightly different opsin polypeptides and, thus, absorb light at different wavelengths. "Red" cones absorb light best around 560 nm, "green" cones around 530 nm, and "blue" cones at 430 nm.[4] Cones allow us to see colors and give much sharper images, but they require brighter light to operate. This is why things appear less colorful at night. Your rods are working, but your cones are not.

Recent discoveries in genetics and phylogenetics suggest that all animal eyes represent a single evolution.[5] This means that eyes in mammals, birds, reptiles, even insects and mollusks share common features and a common genetic inheritance. At the same time, different species have evolved to fit their environment—or simply changed due to random forces. Different eyes see differently. Humans and chimpanzees share the same three types of cones, but most mammals have only two types. Insects share fewer similarities; they use very similar rhodopsins, but have different cell structures for holding them. Butterflies, for example, have rhodopsins that absorb light around 360 nm, 470 nm, and 530 nm. The shorter-wavelength cones allow them to see patterns in what we would consider the ultraviolet range.

Knowledge about different kinds of vision profoundly altered the way I think about physical sentience. Red, green, and blue do not exist in nature. Rather, they are categories assigned by my biology. I see things a certain way because my senses have limited the way I receive input from the outside world. Butterflies see things differently. In fact, many flowers have ultraviolet markings on them at a wavelength no human can see. Butterflies, however, can see and be attracted by them. Differences in sentience shape the way we interact with the world.

How we experience the world becomes particularly relevant when associated with intelligence and reason. The conduct of science relies heavily on sense perception. Observations are only valid if other observers can repeat them. What does it mean for science to recognize the possibility of nonhuman sentient observers? Other observers are almost guaranteed to have a different perspective based on different senses.

The problem is not entirely new to science. We use machines regularly to make measurements we could not make ourselves. We take pictures at ultraviolet and infrared wavelengths. We measure gravity and energy and speed, all quantities not directly available to the five senses. Astrobiologists need to be aware of how our humanity affects our science and how our instrumentation affects our science.

If you think this is a trivial or purely abstract question, consider how you would decide which instruments to put on a spacecraft. Mass and size limits put strict constraints on planetary observing missions. If you can only put 5 kg on the surface of Europa, what instruments—what senses—would you want available?

I have used vision as a brief example of how biochemistry informs both sentience and science. Vision provides yet another example of bottom-up organization as well. The mechanism shows how building blocks can be assembled in new and interesting ways. Evolutionary biologists cannot say exactly how the process happened, but they find compelling similarities between vision and phototrophy. New functions often arise from old mechanisms.[6] The same pieces can be assembled in new ways. The story I tell in this book has shown a gradual increase in complexity, from hydrogen clouds to stars and planets, to elements, to molecules, to cells, and to organisms. Sentience too can be seen as a new possibility, formed from the assembly of items at lower levels.

Emotional Sentience

Physical sentience, although extremely important, cannot give us quite what we want as a defining characteristic for humans. The sensitive soul of Aristotle attributes sentience to all animals. We need something more. The English philosopher Jeremy Bentham (1748–1832) advocated for a moral code based on the promotion of happiness and the minimization of suffering (a form of utilitarianism). Bentham and others have argued that animals suffer and that their suffering—or happiness—should affect our moral reasoning. Along these lines, the term "sentience" is occasionally associated with the ability to feel emotion. Note the ambiguity in English—feeling can refer to touch or to a psychological state. What I have called physical sentience corresponds to touch. What I have called emotional sentience refers to a psychological state.

Recognizing the ability of other organisms to suffer—in the same manner that humans suffer—has a number of results, both abstract and concrete. Buddhism aims at the cessation of all sentient suffering. This implies a particular relationship between humans and other organisms. Buddhism and a number of other Eastern religions advocate for a vegetarian diet; Jains avoid harming any living thing. Some sweep the path in front of them to avoid stepping on insects and eat only fruits and vegetables that can be harvested without killing the plant. Christianity, on the other hand, particularly Protestant Christianity, has leaned toward systems of ethics that only maximize human well-being. This is not the place to judge these moralities, but it is a good time to become aware of how the idea of sentience can have a huge impact on how we view and interact with the biosphere. Do humans have a privileged place? If so, why?

Formal science appears to be unnecessary to address the question of animal suffering. Most of us have experienced relationships with cats, dogs, horses, or other animals that we interpret as happiness and suffering, pleasure and pain. The question of emotional sentience, however, addresses whether or not they have some form of interior life that matches our own. How do they feel? And do their feelings matter?

Sentience has long been tied to animal rights movements, but it also comes up in the context of artificial intelligence. The most famous example might be the Philip K. Dick book *Do Androids Dream of Electric Sheep?* (brought to the movie screen as *Blade Runner*). In the book, Dick explores android (human-shaped robot) feelings and whether those feelings require us to treat them with the same consideration with which we treat humans. Isaac Asimov also delves into this issue repeatedly. Would a robot capable of solving problems, planning ahead, and thinking abstractly be the equivalent of a human? Where do emotions fit into the definition? Does our humanity rest only in intelligence or is there something more, something emotional? Sentience attempts to identify this nonrational element.

Can you have artificial intelligence without artificial life? What would artificial life entail? Should we consider computer programs? Computer scientists have already created programs that replicate themselves with some form of random variation. When placed in environments with limited resources (memory, operating time, or both), they undergo natural selection. In other words, they evolve. Does that mean they are alive?

Our understanding of artificial intelligence cannot be separated from the question of artificial life. We intuitively attach significance to the psychological/physiological state of living things. We use words like visceral, carnal, and incarnational to describe the way people think. Intelligence must be tied up with the metabolism—in our minds at least. As we explore space, astrobiologists will need to be conscious of how this bias affects our definitions and our search for life out there.

Self-Awareness

Another well-known criterion for human individuality addresses the idea of self-awareness. Some philosophers have proposed that humans are unique because we can conceptualize ourselves. The French philosopher René Descartes (1596–1650), who famously remarked "I think therefore I am," suggested that intelligence not only separates humans and all other animals, it defines reality. I cannot imagine a more definitive break between humans and other species. Self-awareness appears to be proof both of abstract thought and the ability to differentiate self from other.

Recent discoveries have put a damper on this train of thought, at least as a defining character for humans. Animal psychologists created a test for self-recognition. They placed animals in front of a mirror and checked to see how they reacted. Most animals assume that the image in the mirror is another member of its own species. It may attempt to communicate, but will not make the cognitive leap required to identify itself in the image. In a few species, individuals demonstrate self-awareness. By means of repetitive motions, they test the mirror to see that it reflects their own actions. They identify themselves in the image. In a further test, the psychologists place a mark on an individual's forehead.[7] Self-aware species will position themselves in front of the mirror in such a way as to identify the mark. They will then touch the mark, having recognized it in the mirror. Using tests like this, psychologists have established that apes, dolphins, and elephants possess a sense of self.

Tool Use

Tool use, both simple and complex, provides some important insights into intelligence. We identified space exploration as one of humanity's signature achievements. Surely, the ability to construct spacecraft and other machines reflects human ingenuity and uniqueness. The ability to use tools was once considered to be the defining characteristic for human intelligence, but in the past century we have begun to appreciate a wide variety of tool use in non-human animals.

Simple tool use can be defined as the ability to use an inanimate object to manipulate other objects. This definition allows us to filter out some actions that may not get at our idea of intelligence. Many species use inanimate objects as dwellings or traps. Nest building in birds and web spinning in spiders may be attributed to instinct. I do not know if that is the best explanation in some cases, but for the moment I would like to have a less ambiguous idea of tool use. The definition above requires some form of planning and a desired outcome.

Even having limited tool use to purposeful activity with an inanimate object, we still see multiple examples in nonhuman animals. Fish, insects, and mammals all use projectiles. They spit water or throw small rocks to catch food or incapacitate an enemy. Otters have been shown to use rocks to crack open the shells of mollusks, and some birds drop turtles to break their shells. Both demonstrate temporal, goal-oriented thinking. Birds and mammals both use sticks to dig insects out of bark and soil. Tool use appears to be quite common.

More complicated versions of tool use have also been seen. Ants demonstrate agricultural proficiency. They are one of a very small number of species that cultivate other species. Some ants show evidence of having kept domesticated fungal species for over 50 million years.[8] Phylogeneticists have reconstructed the evolutionary history of fungal strains and the colonies of ants that cultivate them. The two trees match, demonstrating that the ants and fungi evolved in tandem. More interesting still, a species of fungal parasites—weeds, if you will—show the same tree. Apparently, the ants have been fighting the weeds for a long time. In addition to pest control, ants seed new beds, fertilize them, and prune the fungus.

Ants have also been shown to cultivate other insects. They herd aphids, protecting them from predators, grazing them on leaves, and collecting the honeydew they produce. The association resembles human cattle ranching. Humans cannot digest grass; cattle can, working with the bacteria in their stomachs. Humans herd cattle to transform unusable chemical energy (cellulose) into useable energy (meat and milk). Likewise, ants use aphids and other insects to process the organic molecules in plants. More than half of all butterfly species associate with ants as caterpillars. The caterpillars provide useful excretions and the ants provide protection. Cultivation cannot be limited to humans.

Agriculture demonstrates advanced tool use. This begs the question, "How are humans different, exactly?" Our impact on the environment has been profound and our accomplishments seem unique, but identifying the exact nature of our uniqueness can be problematic. Sentience, self-awareness, and tool use have all failed to distinguish humans.

Intelligence and Life

In the end we find that we have just as much trouble identifying intelligence as we do identifying life. Both concepts are indispensable. At the beginning of this book, I spoke about the earliest researchers, early humans, who defined life by what could be eaten and what could eat us. We still attach high importance to identifying resources, threats, and collaborators. Even at the most basic level, our interactions with the environment depend critically on recognizing life and recognizing intelligence.

Recognizing Life

The second law of thermodynamics tells us that isolated systems tend toward disorder. Any organism wishing to maintain itself will need to identify, use, and, in some cases defend, a source of energy. This means that individuals need to avoid being isolated systems. They need to connect themselves into networks of energy. Upkeep requires attachment to the larger system—a system that ultimately loses energy.

Stars provide the necessary energy to keep planetary biospheres operational. On Earth, individual organisms tap into solar energy directly by using photons or indirectly by getting their energy from other organisms. Phototrophs need only identify the Sun. The rest of us need to identify other organisms. The ability to recognize life is a survival skill because life always contains stored energy. Humans (and all other chemoorganoheterotrophs) get their energy by breaking down other living things. The larger system loses energy—the larger system always loses energy—but the individual survives.

Our fascination with living things can be explained in a number of ways, but it may be an evolutionary necessity. The difference between a rock and a plant may be the difference between life and death if we need to eat. The difference between a rock and a saber-tooth tiger may have been the difference between life and death if it needed to eat. We live by interactions with the larger system; entropy requires it.

A Moral Interlude

At this point, some readers will be drawing moral conclusions. Many readers—and writers—in the past have done so. Alfred Lord Tennyson (1809–1892) struggled with this concept in his poem "In Memoriam." He used the phrase "nature, red in tooth and claw," which came to be associated with Darwinian evolution. Entropy and natural selection are often lumped together with the struggle for survival. People ask whether science reduces all organisms to gladiators, fighting for their lives. Herbert Spencer (1820–1903) coined the phrase "survival of the fittest" to describe Darwin's idea of natural selection. Both phrases lend evolution a rather sinister air. Both contribute to the idea that evolution is a moral proposition—everyone kills to survive.

Curiously, neither "nature, red in tooth and claw" nor "survival of the fittest" came from an evolutionary biologist. "In Memoriam" represents Tennyson's attempt to deal with the death of a friend. Written in 1850, it does struggle with the developing theory of evolution, but comes before Darwin and natural selection hit the scene. Herbert Spencer was an economist. He commented on Darwin, drawing out ideas about how limited resources foster competition. Many in Victorian England believed that God had ordered the world that way to prevent sloth and reward the righteous. Once again, we can see that outside philosophical influences affect the way science is practiced and received. Here economic theory and a Christian theology of divine providence come into play.

Some scientists picked up these ideas about scarcity, but neither entropy nor evolution requires them. On the contrary, I have been trying to emphasize how the tremendous surplus of energy on Earth resulted in hierarchical complexity. One may choose to emphasize competitive interactions (such as predation) or choose to emphasize symbiotic interactions (such as agriculture). Entropy and evolution only force you to recognize that organisms cannot operate as isolated systems. They require energy input.

I digress onto the question of morality because I think it profoundly impacts any discussion of intelligence. The recognition of life and intelligence forms a crucial part of human identity. It determines our behavior toward that which is other. Astrobiology, in asking us to make clearer definitions, challenges us to ask how we treat other organisms and how we would treat nonterrean organisms if we found them.

Recognizing Intelligence

Intelligent life on another planet would force humans to question whether the uniqueness of humanity comes from some particular property or simply reflects our own ego. Intelligence has long been used as a placeholder for that special something that sets us apart. We have tried to define it as sapience, sentience, self-awareness, and tool use, but find that none of these things really limits intelligence to the species *Homo sapiens.*

At the core of many religious (and nonreligious) ethical systems lies the golden rule: do unto others as you would have them do unto you. The key to such a statement comes in how one defines others. Clearly we do not treat bacteria with the same respect as human beings. When it comes to chimpanzees, we get a little more cautious. Utilitarians like Jeremy Bentham argue that we need to treat all sentient animals similarly to humans. Natural law ethicists like Thomas Aquinas argue that a profound gap exists between humans and all other creatures. Drawing the line for intelligence can have a big impact on how we view the world and how we live our lives.

The discovery of intelligent life elsewhere in the universe would force religious humans to reevaluate the meaning of their beliefs. Christians would need to ask whether Jesus Christ died for the sins of all intelligent beings or just humans. Muslims would need to ask whether nonhuman intelligences could act as God's vicegerents or if they would need human supervision in Islam. Hindus would need to ask where alien life fits within the cycle of death and rebirth. Even those without a formal religion would be forced to question whether they would treat the aliens as equals. And these are only a few of the questions. Intelligence informs identity, whether conscious or unconscious. Our self-concept as humans—who we are, what we mean, and where we are going—depends on ideas of intelligence.

The anthropic principle guarantees not only an observer, but an intelligent observer. Astrobiologists know that the philosophical underpinnings of science depend heavily on physical sentience, intelligence, and the knowledge they shape. Science springs directly from the way we observe the universe. We know already that this is not the only way. For many of us, the greatest motivation to find life elsewhere comes from a hope that such life would provide perspective. It would help us step outside of ourselves and discover something fundamental about how we see the world.

SETI

For the past half-century, a program called SETI, the Search for Extraterrestrial Intelligence, has been the icon of this quest. As a research program, SETI seeks to find intelligent ways to scan the local universe for signs of intelligent life. The most common method has been to scan radio frequencies for nonrandom signals arising from outer space. Once local noise (airplanes, human transmissions, and so on) and regular abiological noise (such as pulsars, which produce a metronome-like pulse) have been eliminated, SETI researchers hope to hear a signal from alien intelligences. Radio signals can be produced easily and detected easily. More than that, Earth has been producing them abundantly for over fifty years. This is not the only SETI endeavor, however; other approaches, including optical searches, have been used as well. To date, no compelling evidence has been found.

SETI also describes a particular group of people interested in astrobiology in general. The SETI Institute was founded in 1984 and has a center in Mountain View, California. It employs over a hundred people investigating a broad variety of astrobiological research projects. Their mission is "to explore, understand and explain the origin, nature and prevalence of life in the universe."

Astrobiology as a field has a complicated relationship with SETI. Some astrobiologists feel that searching for intelligence is pointless or problematic. They want to be sure that the general public sees the two endeavors as separate. The search for intelligent life and the study of life as a planetary phenomenon are two very different things. Other astrobiologists recognize that a huge overlap exists between the two pursuits. Everyone recognizes that the SETI Institute produces some of the best research in the field.

Astrobiology and Intelligence

This book has traced the development and organization of life from universal principles to a concrete example, from subatomic particles to the terrean biosphere. At each level, simple units have been combined and recombined to form more and more complex systems. These systems demonstrate complexity in structure, but also in the storage of energy. I have tried to show that intelligence represents another step on the ladder of hierarchical complexity. It allows humans to discern, communicate, and interact in ways that make new things possible.

The ability of humans to send probes into space comes from our ability to communicate with one another. The work of hundreds of scientists and engineers goes into the design and construction of every spacecraft. Thousands more create a field of study that makes such projects possible. Millions contribute to the function of governments with the resources to fund space exploration.

We come now to the end. In the first seven chapters, I introduced the philosophical constructs behind astrobiology. How do we form the questions that make astrobiology possible? What assumptions do we make when we ask our questions? In this chapter, I touch on what it means that we *can* ask questions. We must return to the anthropic principle and recognize the contingency of our situation. Our situation impacts our science and science has become a defining characteristic of our situation. Humanity stands out because of our ability to poke things with sticks and communicate the results.

Don't get me wrong. I would not say that science equals intelligence, but science and technology represent a key example of intelligence in action. Many other examples exist, including government, religion, and art. This, though, is a book on science and on astrobiology. I find it fascinating that the scientific discussion of astrobiology has led us back to the ability to do science.

Life and intelligence both mean "something like us." They mean it at different levels—all intelligence is life. Intelligence means "very much like us," whereas life means only "vaguely like us." Astrobiology pushes these definitions and asks us to question what it would mean for something out there, something alien, to be also "something like us." The answer to that question will have a profound impact on how we see ourselves, how we relate to life on Earth, and how we look at the stars.

20

The Story of Life

The story of life in space can be told as a single narrative—at least the story of life as we know it. It all started with a bang and the rapid expansion of space. What had been a solitary singularity rapidly expanded into the universe. This expansion started 13.5 billion years ago, give or take a billion years, and continues still. Within 500 million years, the dense energy began to cool down enough to form matter. Particles were born. Particles joined together to form atoms. Matter and space became different.

As matter spread out across space, it settled into a pattern. Hydrogen, and occasionally helium, atoms were strewn across ever-widening space in vast clouds. These clouds formed stellar nurseries. Some perturbation of space, perhaps irregularities in the initial event, led to variations in density. Some clouds collapsed under their own gravity. The hydrogen formed into spinning spheres as accumulated angular momentum made the mass move. The spheres flattened out and condensed further into stars and planetary systems. Within the stars, the immense pressure of infalling matter generated enough heat to begin nuclear fusion. For the first time there was location. The furnaces of solar fusion lit up tiny spheres of habitability and order.

Stars did not form alone, but in galaxies and clusters, reflecting the original irregularities of the Big Bang. Stellar fusion generated larger and larger elements, from beryllium to iron. The largest stars exploded into supernovae. Supernovae generated even more elements, spanning the periodic table from lithium to uranium and blowing them out into interstellar space. New clouds formed—still mostly hydrogen, but containing small amounts of all the other elements. These clouds collapsed into new stars and new planets, heavier and with more possibilities.

Four and a half billion years ago, something remarkable happened. Around an unremarkable star, a planet came into being, bearing iron, and carbon, and liquid water abundantly. We do not know if Earth was one among many, unique in some important way, or just lucky. Whatever the case, it was to be the site of life.

In its youngest days, Earth was bombarded by countless meteors. The sky would have been bright with falling stars and the surface hot with the heat of impacts. Rocks large enough to tear through the crust were common. A rock the size of Mars struck Earth, adjusting its composition and ejecting the moon. For the first half-billion years, the planet was a hot, explosive place to be. We call that time the Hadean Eon and life probably started somewhere, sometime amidst the turmoil.

For almost one and a half billion years (~3.9–2.5 billion years ago), Earth looked like other, lifeless planets. During the Archean Eon, one-celled organisms grew, multiplied, and evolved in the oceans. Still, the atmosphere remained reduced, with abundant nitrogen but no available oxygen. Close inspection could have shown that there were already thriving ecosystems at undersea thermal vents or other specialized locations. All in all, though, there would have been nothing that shouted "life" to a casual observer.

Two and a half billion years ago, all of that changed. One species, perhaps a close relative to the modern cyanobacteria, evolved oxygenic photosynthesis. The species enabled life to colonize every corner of the planet. The biosphere expanded and pumped out massive amounts of oxygen, changing the planet forever. The atmosphere became oxidized, poisoning most living species and paving the way for multicellular organisms. Ozone accumulated in the upper atmosphere, blocking dangerous UV rays and making life on land feasible. The Proterozoic Eon had begun.

During the Proterozoic, more complex forms of life began to emerge. Cyanobacteria and proteobacteria began living inside the cells of eukaryotes, leading to mitochondria and chloroplasts. The first multicellular plants and animals developed; it was not until the Phanerozoic Eon, however, that these forms truly took off. In the wake of a tremendous ice age, the Phanerozoic was launched 542 million years ago. It marks the period of massive multicellular life and fossils. Plants, animals, and fungi colonized the land. The forms with which we are most familiar, mammals and flowering plants, came along hundreds of millions of years later.

Humans only entered the scene 5–10 million years ago. Genetically and biochemically, we closely resemble chimpanzees, but we developed a curious kind of intelligence that sets us apart from other species. In the past two hundred years our knowledge and technology have brought us the power to dominate most other species on the planet. We have the ability to visit other planets and ask questions about origins and meaning.

The earliest life on Earth would have been simple. We find it easier to step from nothing to simplicity than from nothing to complexity. Although we do not know exactly how life started, it seems reasonable to assume that it started simply. Certainly the oldest organisms are small, and though we can only guess at their biochemistry, it too must have been less convoluted than life today. Overall, life remains simple. Bacteria and Archaea still comprise the vast majority of life on Earth. One-celled organisms dominate the biosphere. They live in places no multicelled organism could go, but they have also colonized the surfaces of multicellular life. We all have billions of microorganisms living on our skin and in our gut, defending us from parasites, helping digest our food, and protecting us from all manner of harm. Even simpler than bacteria, viruses and selfish genetic elements float around, evolving with nothing more than a strand of nucleotides to call their own.

We see the world as developing complexity because we look for complexity. We value complexity. And so we tell the story from the beginning of time as the story of how we came to be. As humans, we tell the story in the context of our humanity. On the other hand, evolutionary biologists attempt to see humanity, even intelligence, as a rare, albeit interesting, side note in the history of life on Earth. Humans represent a strange experiment in the use of neurological function. Only time will tell whether this experiment will prove fruitful or simply fade away, as have other experiments. If all the humans suddenly disappeared, the rest of the biosphere would continue quite smoothly on its way.

This leaves astrobiologists in a quandary. As humans we recognize our own importance. As scientists we cannot help but notice how contingent and unremarkable we seem to be in the context of life. We are one species among many. Although we seek to define intelligence, increasingly we realize that what we really define is what it means to be human.

As with any good story, the events may be relayed in a number of ways. The story lies as much in the telling as in the facts. By presenting the information differently, we can begin to see both the facts and their significance in a new light. The characters you choose to follow impact the shape the story takes. In Chapter 7, I characterized life in the cosmos in terms of three particulars. Life must have a setting in which it takes place, a substance from which it is constructed, and an energy currency through which it does work. These three—medium, matter, and energy—form the basis of life as we understand it. In our case, they are represented by water, carbon, and electrons. The next three sections will trace them from the beginning to the end of our story.

Follow the Water

Water can be traced the most easily. It occurs abundantly in our own Solar System and probably throughout the universe. Water forms from two hydrogen atoms covalently bonded to an oxygen. The hydrogen has been around from the beginning, but oxygen must have formed within a star.

This means that water-bearing stellar systems will be secondary systems. They must form from the debris of other stars. The inner Solar System is mostly made up of rock, but the outer Solar System has abundant water-ice. The real question in the history of terrean water arises when we ask how water came to be so common on Earth.

Life probably developed in an aqueous environment on Earth. Cells, as we know them, can only occur in water. At the same time, cells and membranes clearly evolved to separate, contain, and mix aqueous solutions. Terrestrial (land-dwelling) organisms had to learn how to carry and protect their water. Aquatic organisms had to develop tricks to balance the concentrations of molecules inside and outside the cell.

Water remains a limiting factor. Organisms have evolved that are capable of surviving extremes of temperature and pressure, vacuum and toxic chemicals, even the effects of massive radiation. All remain quite sensitive to water loss. Astrobiologists who look for life elsewhere know that water will be a key indicator for life as we know it. Liquid water with a good mix of ions and carbon molecules forms the foundation of terrean life.

Follow the Carbon

Carbon will be a little harder to trace, though the story starts the same way. Carbon, like oxygen, forms within stellar furnaces. A stellar system with carbon must be at least a second-generation system. In our case, carbon needed to be present in the formation of the Solar System. Carbon does not form a major component of any planets, but it does occur abundantly in the atmospheres of Venus and Mars in the form of carbon dioxide. Earth's atmosphere may have had large quantities of this gas before the oxygenic revolution. In the outer Solar System, Uranus, Neptune, and Titan (a moon of Saturn) all have atmospheres high in methane. So carbon is generally available, if not in surplus.

Huge reservoirs of carbon can be found within the biosphere, but they must have accumulated gradually from the earliest days of life on Earth.

Perhaps the process began with self-replicating biomolecules—RNA enzymes or information-storing proteins. Perhaps it started with a cluster of lipid vesicles. We may never know. We do know that wherever the process started, within a billion years it had developed into something we would recognize.

All terrean life follows the same basic pattern. Organisms store information within sequences of nucleic acids, DNA and RNA. Ribosomes manifest this information in the form of polypeptides. Polypeptides do work in the system. Organisms constantly construct and deconstruct organic molecules. They produce or store energy, create necessary molecules, and break up wastes. Carbohydrates store energy and act as structural elements while lipids form membranes to separate solutions and support reactions. We know all terrean life shares a common origin because of this tremendous similarity in metabolism. Even those entities that forgo a cell membrane—viruses and their kind—use the machinery of cellular organisms to reproduce. Even they process information in the same fashion. Biomolecules form one level of organization within all living matter. Carbon enters the system by way of autotrophs. Autotrophs string together carbon atoms acquired from carbon dioxide, methane, and other one-carbon molecules. Heterotrophs get their carbon by eating the autotrophs or feeding off of the organic molecules they produce. Biomolecules can be assembled like building blocks to form larger and larger structures, from organelles to cells, from cells to tissues and organs, all the way up to organisms and ecosystems. At every level, the number of carbon molecules grows and the energy needed to support a common whole increases.

Follow the Energy

Energy will be the hardest of the three essentials to follow. Perhaps this comes from the fact that it is hardest to visualize. Perhaps it comes from the ubiquity of energy in the universe. And, perhaps, it arises from how energy, information, and complexity relate to one another. I cannot say. I do know that many of the most interesting questions—questions having to do with what it means to be alive—arise in the context of energy. We do not find it intuitive, but we do find it important.

The energy story starts at the beginning—the Big Bang. To some extent, the second law of thermodynamics implies that everything after that runs downhill. All energy, order, and complexity in the universe come from the original explosion. The heat of that one event powers everything we do. If it helps, you might consider that everything was packed very neatly into a bundle smaller than an atom—just like a wrapped deck of cards. As space spreads out, the disorder increases. One tiny, brilliantly hot moment gradually loosens. Some tiny points within the vastness of space heat up. All of space cools down. There is something depressing about the thought of disrupting such perfect order, but without unwrapping the deck, you cannot play the game.

Stars represent local furnaces, burning up a huge supply of fuel and producing heat, light, and new elements. Settled in close to one such furnace, Earth absorbs energy from its star. The atmosphere traps some of the energy as heat. Greenhouse gases, like those on Venus, catch light reflected off of the surface and raise the local temperature. This alone is not unique. Many planets must do the same thing. The drama arises when organic chemistry develops that can catch and store energy within carbon compounds and metabolic systems.

All known organisms store both potential energy and chemical energy. Phototrophs use the energy from photons to charge electrons. Oxidation/ reduction (redox) reactions transfer electrons from one molecule to the other. A chain of redox reactions within phototrophs shuttles protons across a membrane, building up an osmotic potential. The osmotic potential can be used to form ATP, the basic unit of energy currency in the cell. The redox chain, meanwhile, often ends by charging a molecule of NADH or NADPH, which can do chemical work or produce more ATP.

All this represents the most basic level of energy storage and use within a cell. It serves to power the simplest metabolism. Life can do more, though. More complex storage devices can be formed in larger and larger molecules. Carbohydrates and lipids store energy in a fashion that can be used later, or by different organisms. In terrean metabolism, glucose may be the most important of these molecules. Heterotrophs, such as humans, break down the glucose, forming more ATP to use in their own cells.

Interactions between organisms can be seen as the transfer of energy to support a genetic agenda. DNA forms the core; a replicating molecule within each replicating organism. If we want to look at information, most of the information resides in the DNA. The other machinery (proteins, lipids, metabolism, and so on) exists for the benefit of replicating genes. Each entity duplicates, maintains, and spreads its genetic material. At the macroscopic level, species interactions match this pattern. Predation—killing and eating prey—constitutes the appropriation of resources for the survival of the predator's DNA. The decomposition of organisms involves opportunistic scavenging for resources from dead organisms. Even pathogens, some of which do not kill the host, look an awful lot like genes taking advantage of foreign metabolism. At the microscopic level, selfish genetic elements appropriate the metabolism of their host to create copies of themselves. Viruses have little more purpose than to propagate genes. When viewed through this filter, evolutionary biology becomes quite clear. Evolution does not mean that organisms try to compete. It means that the genes passed on begin to dominate. They must, because the other genes have ceased to exist.

Intelligence may present an opportunity for more thoughtful pursuit of genetic selection. Humans have the ability to defend their gene lines through choice as well as habit. For this reason, we have been tremendously successful at defending our species against predators and pathogens. (Curiously, we have a tendency to defend human bodies from scavengers as well.) Humans will also defend themselves and their families from more distantly related humans. The love of kin and country could be explained easily in terms of preserving genes. After all, the more closely related a person is, the more of your genes they will share. You can see how the ability to identify others similar to yourself could become important very quickly.

On the other hand, intelligence also acts against the genetic imperative. Humans demonstrate self-sacrificing behavior. Sometimes this will be in the benefit of a large genetic group—giving up self for family, or family for country—but often humans sacrifice self for the sake of an idea or an ideal. These phenomena lead some evolutionary biologists to believe that intelligence supports a whole new level of energy storage. Intelligence allows humans to create narratives, theories, and beliefs. We create books and CDs and DVDs to store information in the form of words and sounds independent of a biological entity. We make batteries and machines to store energy independent of a metabolism.

Is intelligence yet another step up on the ladder of complexity? Can it be separated from the lower levels? Right now, we cannot say, but astrobiologists deal with this question as well. It has something to do with how we understand our own place on the planet and our place in the universe. With such a broad and amorphous topic—life on Earth—we must restrain ourselves to definite and answerable questions.

Astrobiology remains a nascent science. For now, astrobiologists create questions that can be answered in a meaningful way by science. We investigate the origins of energy in the universe. We look at the formation of stars and planets. We pry into the beginnings of life on Earth and watch how terran life develops. At the same time, we explore possibilities of life elsewhere. We ask how stars, planets, and life are related, and theorize how that might apply elsewhere.

Two important assumptions reside at the center of this endeavor. First, the formation of stars and the formation of life can never be fully separated. In order to understand life on Earth, we must view it in context. Second, to answer the most profound questions, scientists from a variety of disciplines need to learn to communicate. The NASA Astrobiology

Roadmap, a document that outlines research goals and objectives to further NASA's strategic planning efforts, identifies three important questions: "How does life begin and evolve?" "Does life exist elsewhere in the universe?" "What is the future of life on Earth and beyond?"[1] These questions are too big to tackle with a single research program, but they inspire individual projects that will take small steps toward a larger goal—the goal of understanding ourselves and our place in the universe. The study of life in space will require cross-fertilization from many fields of knowledge—cosmology, physics, astronomy, chemistry, biology, and information science. It may even draw on philosophy, psychology, computer science, and other fields. We stand at the very beginning of a very big project.

Among scholars, the project should not be limited to science; it should include certain fields in the humanities as well. The questions are too big, and the results are too important. Neither can the science be corrupted by the goals of religion and ideology. Astrobiologists must establish the goals and acknowledge the biases of science so that it can be used as a tool in a larger endeavor. Society as a whole will have to evaluate the impact of our growing knowledge of self and cosmos.

In the story of life I tell in this book, I have mostly limited myself to science, though I admit I have dropped a few philosophical hints on the way. The conclusions are my own, but I hope they will tell you something interesting about the field of astrobiology and about life in space. In closing, I hope you will take away at least four important ideas about life. They have been key to my own understanding.

First, we have exactly one example of life in space—terrean life. We need to think carefully about how this impacts our understanding. Our own existence demonstrates a deep interconnectedness of physics, chemistry, geology, and biology.

Second, the line between what we know and what we do not know can be hard to see. Astrobiology involves questioning how we arrive at answers as well as the answers themselves. Life, intelligence, and individuality all represent useful, intuitive categories, but our definitions of them can lead us astray.

LIFE IN SPACE

Third, life depends upon identifying and interacting with other life. Entropy forces us to live by working within a larger biological system.

Finally, our biochemistry impacts our identity, science, and our beliefs. It may not determine them, but it does have an impact. Our intelligence appears to have novel properties unique to such a complex level of organization. We must be ever vigilant to how our biology affects our morality.

If astrobiology could be said to have one underlying message, it would be that life is a great deal more confusing and wonderful than anyone suspected. At the same time, we have learned that all life—at least all life on Earth—shares some remarkable similarities. Astrobiology straddles the divide between what we know and what we do not, between life as we know it and the possibility of something else. It cannot begin, or end, in anything but wonder at the immensity of the questions. What is life, what does it mean, and what is our place in it all?

ABBREVIATIONS

NOTES

ACKNOWLEDGMENTS

INDEX

Abbreviations

ADP	adenine diphosphate
AIDS	acquired immune deficiency syndrome
AMP	adenine monophosphate
ATP	adenine triphosphate
AU	astronomical unit
CHNOPS	carbon, hydrogen, nitrogen, oxygen, phosphorus, and sulfur
CoA	coenzyme A
DNA	deoxyribonucleic acid
ESA	European Space Agency
$FADH_2$/FAD	flavin adenine dinucleotide
GTA	gene transfer agent
HIV	human immunodeficiency virus
IAU	International Astronomical Union
JAXA	Japan Aerospace Exploration Agency
JPL	Jet Propulsion Laboratory
KBO	Kuiper belt object
MESSENGER	*Mercury Surface, Space Environment, Geochemistry, and Ranging*
MHC	major histocompatibility complex
mRNA	messenger RNA
NADH/NAD	nicotinamide adenine dinucleotide

NADPH/ NADP	nicotinamide adenine dinucleotide phosphate
NASA	National Aeronautics and Space Administration
NEA	near-Earth asteroid
NEAR	*Near-Earth Asteroid Rendezvous*
OGLE	Optical Gravitational Lensing Experiment
PAH	polyaromatic hydrocarbon
RNA	ribonucleic acid
rRNA	ribosomal RNA
RV	radial velocity
SETI	Search for Extraterrestrial Intelligence
SSU	small subunit (rRNA)
TCA	tricarboxylic acid
TNO	trans-neptunian object
TNT	trinitrotoluene
tRNA	transfer RNA
UV	ultraviolet
WMAP	Wilkinson Microwave Anisotropy Probe

Notes

1. Caught Up in Life

1. In planetary science, terrestrial means Earth-like and describes rocky planets such as Mars as opposed to gas planets such as Jupiter. In botany, terrestrial refers to plants growing on land.

2. One of the best recent attempts to cover the field in a rigorous way is W. T. Sullivan III and J. A. Baross, eds., *Planets and Life: The Emerging Science of Astrobiology* (Cambridge: Cambridge University Press, 2007).

2. Living Science

1. I became interested in writing this book while trying to solve that very problem. Along with almost a hundred other scientists, I put together the "Astrobiology Primer," aimed at bridging the gaps between different fields. It contains a very dense introduction to astrobiology with extensive references. L. J. Mix, J. C. Armstrong, A. M. Mandell, A. C. Mosier, J. Raymond, S. N. Raymond, F. J. Stewart, K. von Braun, and O. Zhaxybayeva, eds., "The Astrobiology Primer: An Outline of General Knowledge—Version 1, 2006," *Astrobiology* 6(2006): 735–813. Mary Ann Liebert Publishing has it available for download at http://www.liebertonline.com/toc/ast/6/5.

2. Actually, they can break down at black holes, quantum singularities, and the beginning of the universe, but they work for the vast majority of space-time. The idea that the rules apply constantly through space and time will be dealt with in Chapter 4.

3. They will each decay by beta radiation, emitting an electron and an antineutrino. This forms the basis of radiocarbon dating.

4. The reason this experiment does not work outside a vacuum tube involves air resistance. The shape of an object will affect how well it displaces air on the way down and the feather will slow noticeably.

5. This is not to say that a "life essence" does not exist, simply that one is not necessary to explain inheritance. The change in thought rests notably with the work of Charles Darwin (evolution, 1859), Gregor Mendel (genes, 1867), Theodosius Dobzhansky (speciation, 1937), and James Watson and Francis Crick (the structure of DNA, 1953).

6. Notable scientists include Dmitri Mendeleev (periodic table of elements, 1869), Ernest Rutherford (protons, 1918), and James Chadwick (neutrons, 1932).

7. D. S. McKay, E. G. Gibson Jr., K. L. Thomas-Keprta, H. Vali, C. S. Romanek, S. J. Clement, Z. D. F. Chillier, C. R. Maechling, and R. N. Zare, "Search for life on Mars: Possible relic biogenic activity in Martian meteorite ALH84001," *Science* 273(1996): 924–930.

3. Defining Life

1. Jacobellis v. Ohio, 378 US 184 (1964), Justice Stewart concurring.

2. Paris Adult Theatre I v. Slaton, 413 US 49 (1973), Justice Brennan dissenting.

3. For more details see: J. D. Farmer and D. J. Des Marais, "Exploring for a record of ancient Martian life," *Journal of Geophysical Research* 104(1999): 26977–26996.

4. S. J. Mojzsis, G. Arrhenius, K. D. McKeegan, T. M. Harrison, A. P. Nutman, and C. R. L. Friend, "Evidence for life on Earth before 3,800 million years ago," *Nature* 384(1996): 55–59.

5. J. W. Schopf, "Microfossils of the early Archean apex chert—New evidence of the antiquity of life," *Science* 260(1993): 640–646.

6. In my limited expertise, I think that Mojzsis biomarkers will win out, but the case against can be found in: M. A. van Zuilen, A. Lepland, and G. Arrhenius, "Reassessing the evidence for the earliest traces of life," *Nature* 418(2002): 627–630. On the other hand, I'm beginning to doubt Schopf's fossils based on the arguments found in: M. D. Brasier, O. R. Green, A. P. Jephcoat, A. K. Kleppe, and M. J. van Kranen-

donk, "Questioning the evidence for Earth's oldest fossils," *Nature* 416(2002): 76–81. In both cases, however, the final verdict will come from more research and an eventual consensus.

7. Some chemists limit organic chemistry to molecules with carbon-carbon bonds. Others limit organic chemistry to molecules with carbon-hydrogen bonds. I will use the former definition as I find it most helpful.

8. Entropy gets a fuller treatment in the next chapter.

9. It should be noted, however, that some simple plants, fungi, and animals (sponges) can be mistaken for mineral growths and vice versa. Paleontologists learn early on to test carefully and look for minute differences.

10. The most extreme example is *Deinococcus radiodurans,* a bacterium that can withstand desiccation, toxic chemicals, and ridiculous levels of radiation by spell-checking its own DNA. *D. radiodurans* expends high levels of energy, but can exist in environments that would liquefy other organisms.

11. A good review of so-called "horizontal gene transfer" dealing particularly with the impact on tracing evolution can be found in: J. P. Gogarten, W. F. Doolittle, J. G. Lawrence, "Prokaryotic evolution in light of gene transfer," *Molecular Biology and Evolution* 19(2002): 2226–2238. For the recent discovery of "gene transfer agents" in *Proteobacteria,* see: A. S. Lang and J. T. Beatty, "Genetic analysis of a bacterial genetic exchange element: The gene transfer agent of *Rhodobacter capsulatus,*" *Proceedings of the National Academy of Sciences USA* 97(1999): 859–864.

12. Harold Morowitz has written on the topic of emergence, most recently *The Emergence of Everything: How the World Became Complex* (New York: Oxford University Press, 2002). Although I do not follow his reasoning nor agree with all his assumptions, he presents an interesting analysis.

13. Exploration of the biochemical possibilities has been quite rigorous, and it would be impossible to even survey the literature here, though there is a survey in "The Astrobiology Primer." I touch on the topic again briefly in Chapter 18, but let me mention a few names. The "RNA-world" with RNA-only life: W. Gilbert, "The RNA world," *Nature* 319(1986): 618. The "protein world": J. H. McClendon, "The origin of life," *Earth-Science Reviews* 47(1999): 71–93. The "lipid world": D. Segre, D. Ben-Eli, D. W. Deamer, and D. Lancet, "The lipid world," *Origins of Life and Evolution of the Biosphere* 31(2001): 119–145. Cells prior to biochemistry: J. W. Szostak, D. P. Bartel, and P. L. Luisi, "Synthesizing life," *Nature* 409(2001): 387–390. Biochemistry first: F. A. Anet, "The place of metabolism in the origin of life," *Current Opinion in Chemical Biology* 8(2004): 654–659.

14. C. Cleland and C. Chyba, "Defining 'life,'" *Origins of Life and Evolution of the Biosphere* 32(2002): 387–393.

4. A Well-Behaved Universe

1. Immanuel Kant, *Critique of Pure Reason* (1781). Introduction, Section V.2.

2. In practice, physicists have become fairly confident about converting matter into energy. Nuclear bombs do this quite efficiently. Converting energy into matter appears to be a more difficult problem.

3. It so happens that there are other kinds of particles and even protons, neutrons, and electrons can be broken up into smaller units such as baryons, leptons, and quarks. To learn more about these, read Kenneth Ford, *The Quantum World* (Cambridge, MA: Harvard University Press, 2004).

4. This scheme, too, can be reduced. Again, read Ford, *The Quantum World,* for more details.

5. All objects obey the common formula $F = Gm_1m_2/r^2$ where F is a force of attraction, G is the gravitational constant (6.67×10^{-11} Nm^2kg^{-2}), m_1 and m_2 are the masses of two objects, and r is the distance between their centers of mass.

6. All objects obey the common formula $F = ma$ where F is any force, m is the mass of the object, and a is its acceleration (or change in velocity) as it travels through space.

7. This happens because r is larger. The effect is extremely small because relative to the diameter of Earth, the altitudes humans travel are minuscule.

8. This results from the lower air pressure at higher altitudes—also a factor of gravity. Water boils when the vapor pressure (a function of temperature) equals the air pressure. In interstellar space, pressure is so low that water will turn into a gas immediately.

9. For an easily accessible look at relativity, I would recommend Larry Gonick and Art Huffman, *The Cartoon Guide to Physics* (New York: Harper Collins, 1991). It is fun for all ages. More intense readers may enjoy E. F. Taylor and J. A. Wheeler, *Spacetime Physics* (New York: W.H. Freeman, 1992).

10. The verbs "rotate" and "revolve" are often used interchangeably. Astronomers distinguish movement of an orbiting object—revolution—from the movement of a spinning object—rotation.

11. This is true for large movements. For smaller movements, the problem can be quite tricky. Try laying out a pattern on a table in front of a mirror. Then trace the pat-

tern with a fingertip while only looking at the reflected image. The constant conversions (related to transformation of the image by reflection) can be quite challenging.

12. Redshift occurs when a light source moves away from the observer. If you have heard a siren traveling away from you, you may have noticed that the sound changes. The frequency gets lower and lower because the waves get stretched out. For sound waves this is called a Doppler shift, but the same pattern can be found for light waves when the light source is moving relative to the observer. For light, the process is called redshift—or blueshift when the source and observer are moving toward one another.

13. E. Hubble, "A relation between distance and radial velocity among extragalactic nebulae," *Proceedings of the National Academy of Sciences USA* 15(1929): 168–173.

14. That is, 71 (+0.04/-0.03) kilometers per second per megaparsec (a megaparsec being about 3.26 million light years). For recent calculations of the Hubble constant based on cosmic microwave background radiation, see map.gsfc.nasa.gov/m_uni/uni_101expand.html and the related article: D. N. Spergel, M. Bolte, and W. Freedman, "The age of the universe," *Proceedings of the National Academy of Sciences USA* 94(1997): 6579–6584.

15. NASA maintains an excellent Web site that explains this and many other topics at map.gsfc.nasa.gov/m_uni/uni_101bbtest3.html.

16. WMAP is not the only mission of significance. The Cosmic Background Explorer satellite (COBE, launched 1989) contributed the data for the original discovery that background radiation was not uniform. Caltech has made discoveries in this area with Boomerang—a balloon-supported platform—and the European Space Agency plans to launch the *Planck* spacecraft in 2008.

17. Macroscopic examples are often slightly imperfect. In this case it should be noted that the wind is an outside force. Gravity was acting from the beginning, so that does not count.

18. Entropy usually indicates the measure of disorder within a system. Properly the entropy of a system and its environment increases or stays the same for any process.

19. All atoms you encounter are heated. Super-cooled particles have only recently been generated in laboratories. Even so, no one has been able to cool a particle to absolute zero (0 K or $-273.15°C$). Even in space, particles have a few degrees of heat.

20. The classic treatment of this can be found in: Stephen Hawking, *A Brief History of Time* (New York: Bantam, 1996).

5. Well-Behaved Observers?

1. Some authors prefer the term "selection effect." Note that in statistics, bias does not indicate an unreasonable predisposition or prejudice. Rather, statisticians use the word "bias" for a systematic error that results, usually unintentionally, from the way the question was asked, the hypothesis framed, or the experiment carried out.

2. Brandon Carter, "Large number of coincidences and the anthropic principle in cosmology," *Proceedings of the IAU Symposium 63* (Dordrecht: Reidel, 1974), 291–298.

3. Brandon Carter, "The anthropic principle and its implications for biological evolution," *Philosophical Transactions of the Royal Society of London, Series A* 310(1983): 347–363.

4. Francis Bacon, *Novum Organum.* Aphorism XLVI, in particular, deals with the tendency of those saved from a shipwreck to attribute their salvation to divine intervention without asking whether divine intervention also caused the drowning of their comrades. Survivors construct a plan to explain their survival, whereas the dead do not construct plans.

5. J. D. Barrow and F. J. Tipler, *The Anthropic Principle* (New York: Oxford University Press, 1986).

6. Nick Bostrom, *Anthropic Bias* (New York: Routledge, 2002).

7. To learn more about the fundamental forces, read Kenneth Ford, *The Quantum World* (Cambridge, MA: Harvard University Press, 2004).

8. Hugh Everett (1930–1982) originally proposed the many-worlds interpretation in 1957. Hugh Everett, "Relative state formulation of quantum mechanics," *Review of Modern Physics* (1957): 454–462. It was later developed by Bryce Seligman De Witt (1923–2004). Bryce De Witt, "Quantum mechanics and reality," *Physics Today* 23, 9(1970): 30–35.

9. Plato and Thomas Aquinas are two famous examples.

10. Phil Dowe presents an interesting exploration of these questions in the context of philosophy in religion and science. See especially chapters 6 and 7. Phil Dowe, *Galileo, Darwin, and Hawking* (Grand Rapids, MI: Eerdmans, 2005).

11. Bayesian theory can be encountered in a basic way in elementary statistics textbooks. Those wishing a more rigorous introduction to philosophy of science should look at Colin Howson and Peter Urbach, *Scientific Reasoning: The Bayesian Approach* (Chicago, Open Court: 2005).

6. Life in the Cosmos

1. The known limit appears to be 60 percent relative humidity; see G. M. Marion, C. H. Fritsen, H. Eiken, and M. C. Payne, "The search for life on Europa: Limiting environmental factors, potential habitats, and Earth analogues," *Astrobiology* 3(2003): 785–811.

2. Technically, the oxygen has an electronegativity of 3.5 while hydrogen only has an electronegativity of 2.1. The higher the electronegativity, the stronger the attraction. For a good introduction to chemistry that is both basic and fun, check out Larry Gonick and Craig Criddle, *The Cartoon Guide to Chemistry* (New York: Harper Collins, 2005).

3. P. D. Ward and S. A. Benner, "Alien biochemistries" in W. T. Sullivan and J. A. Baross, eds., *Planets and Life: The Emerging Science of Astrobiology* (Cambridge: Cambridge University Press, 2007), pp. 537–544. See section 27.3, pp. 539–541.

4. Ionic bonds form when a positively charged atom associates with a negatively charged atom.

5. William Bains, "Many chemistries could be used to build living systems," *Astrobiology* 4(2004): 137–167.

6. Some prominent examples include coral and stromatolites, organisms that deposit a rocklike structure which slowly accumulates under a layer of living cells.

7. Specifically: $CH_3COOH + NaHCO_3 \rightarrow CH_3COONa + H_2CO_3\ H_2CO_3 \rightarrow H_2O + CO_2$.

8. Incidentally, in many reactions this process accompanies the addition of an oxygen atom, hence "oxidation."

9. Likewise, the word "reduction" relates to the total amount of oxygen in a molecule being reduced. Although the original definition of the words comes from the gain and loss of oxygen, modern chemists generally prefer to think of the loss and gain of electrons. It makes for a more comprehensive definition, if slightly counterintuitive to the nonscientist.

10. Although we conventionally place the positive charge on the nitrogen because of the number and quality of the bonds it forms, to some extent the charge can be said to be distributed around the ring. If this seems confusing, don't worry. Many scientists find it confusing, too. Tracking electrons in large molecules can be difficult, if not impossible. For more information, look up "electron resonance" and "aromatic chemistry" in any organic chemistry or physical chemistry textbook.

11. Exact measurements are difficult; however, NADH has roughly as much reducing power as 2.5–3 ATP, and NADPH has the power of 3.5–4 ATP.

12. The process resembles the way smoke will spread out in an enclosed room or ink will disperse in water. The increase in entropy, or movement toward maximum disorder, means that any substance gathered together in one place will tend to move outward until evenly distributed.

7. Life among the Stars

1. A quick and dirty introduction to stars in astrobiology can be found in K. von Braun and S. N. Raymond, "Stellar formation and evolution," in L. J. Mix, J. C. Armstrong, A. M. Mandell, A. C. Mosier, J. Raymond, S. N. Raymond, F. J. Stewart, K. von Braun, and O. Zhaxybayeva, eds., "The Astrobiology Primer: An Outline of General Knowledge—Version 1, 2006," *Astrobiology* 6(2006): 741–748. A more extensive description of stars can be found in R. A. Freedman and W. J. Kauffmann III, *Universe* (New York: W. H. Freeman and Co., 2005).

2. A very popular book addressing the origin of the universe is Stephen Hawking, *A Brief History of Time* (New York: Bantam Books, 1998).

3. Astronomers often measure temperature in the Kelvin scale. It is useful when the temperature approaches absolute zero, the temperature at which everything freezes (0 K, −273°C, −460°F). Centigrade and Fahrenheit scales can be most useful around the freezing point of water (0°C, 32°F). Conversions can be made easily with the following formulas: (9/5)°C + 32 = °F, K + 273 = °C. Because the Kelvin scale is absolute, it uses no degrees. In general, I will use the familiar Fahrenheit scale when talking about habitable and potentially habitable environments. Readers might be interested to know that even the space between stars appears to be warmer than absolute zero. Small vibrations of the rare hydrogen atom and the passage of light through space heat it up slightly. Current estimates of interstellar space suggest a temperature near 2.7 K.

4. For a detailed discussion, check out M. Claire, "Global Climate Evolution," in Mix et al., "The Astrobiology Primer," 760–765.

5. Remember that red light occurs at the long-wavelength, low-energy range of the visible spectrum. Cooler, less energetic light is redder light.

6. The second law of thermodynamics still holds, however. For all the energy that is generated, more energy gets lost in the process. Stars collect the potential energy of countless atoms drifting in space and concentrate it. At the end of the process the universe is colder and more diffuse than at the beginning.

7. D. F. Figer, "An upper limit to the masses of stars," *Nature* 434(2005): 192–194.

8. Potential energy from the strong nuclear force is converted into radiant energy.

9. Kinetic energy is converted to potential energy from the electric force.

10. Neutrinos are subatomic particles with no charge and negligible mass. Neutrinos carry the extra energy away when the proton and electron collide.

11. Greater than 4×10^{17} kg/m^3.

12. Infalling particles can accelerate to 15 percent of the speed of light.

13. A. L. Melott, B. S. Lieberman, C. M. Laird, L. D. Martin, M. V. Medvedev, B. C. Thomas, J. K. Cannizo, N. Gehrels, and C. H. Jackman, "Did a gamma-ray burst initiate the Late Ordovician mass extinction?" *International Journal of Astrobiology* 3(2004): 55–61.

14. The largest mass extinction occurred at the end of the Cretaceous period. This event marks the line between the Cretaceous and Tertiary periods in paleontology. Scientists call it the K-T event, based on the abbreviations for the periods. Astrobiologists believe that a giant meteor impact eliminated 85 percent of all terrean species at this time.

15. As it turns out, this ozone results from oxygenic photosynthesis, meaning that life contributes to the life-friendly environment here on Earth. Chapter 12 will deal briefly with the rise of ozone on Earth and how that affected the development of life. Chapter 16 will cover phototrophy and how life turns light at certain wavelengths into chemical energy.

16. For the purposes of astrobiology, chemically interesting means metabolic reactions.

17. The visible spectrum covers a range of 400–700 nanometers (nm). Light that has slightly shorter wavelengths is called ultraviolet because it lies beyond dark blue, whereas slightly longer wavelengths of light are called infrared because they lie beyond red.

8. The Planetary Phenomenon

1. A quick introduction to planet formation in astrobiology can be found in S. N. Raymond, "Planet formation and dynamical evolution," in L. J. Mix, J. C. Armstrong, A. M. Mandell, A. C. Mosier, J. Raymond, S. N. Raymond, F. J. Stewart, K. von Braun, and O. Zhaxybayeva, eds., "The Astrobiology Primer: An Outline of General Knowledge—Version 1, 2006," *Astrobiology* 6(2006): 748–752.

2. I can only give a cursory overview here. Readers interested in knowing the full details are encouraged to read J. E. Chambers, "Planetary accretion in the inner Solar System," *Earth and Planetary Science Letters* 223(2004): 241–252. More recent research on terrestrial planet formation has suggested faster accretion. See M. R. Meyer, J. M. Carpenter, E. E. Mamajek, L. A. Hillenbrand, D. Hollenbach, A. Moro-Martin, J. S. Kim, M. D. Silverstone, J. Najita, D. C. Hines, I. Pascucci, J. R. Stauffer, J. Bouwman, and D. E. Backman, "Evolution of mid-infrared excess around Sun-like stars: Constraints on models of terrestrial planet formation," *Astrophysical Journal* 673(2008): L181–184.

3. Further information on both models can be found in A. P. Boss, "Formation of gas and ice giant planets," *Earth and Planetary Science Letters* 202(2002): 513–523.

4. Raymond, "Planet formation and dynamical evolution," p. 750.

5. Figure 8.1 is based on parameterizations derived from J. F. Kasting, D. P. Whitmore, and R. T. Reynolds, "Habitable zones around main sequence stars," *Icarus* 101(1993): 108–128. The parameters were designed to explain stars in the 3,700–7,200 K range and may not be as useful for hotter and cooler stars. The tapered effect on the top of the diagram may be an artifact; however, hotter stars lead shorter lives and thus are poor candidates in a search for life.

6. Sean Raymond, an astrobiologist at the University of Colorado, has reminded me that near-circular orbits may not be necessary for life. Apparently, I have some Earth-centric biases of my own. Interested readers might take a look at F. Selsis, J. F. Kasting, J. Paillet, and X. Delfosse, "Habitable planets around the star Gl581?" *Astronomy and Astrophysics* 476(2007):1373–1387. Also, see D. M. Williams and D. Pollard, "Earth-like worlds on eccentric orbits: Excursions beyond the habitable zone," *International Journal of Astrobiology* 1(2002): 61–69.

9. The Inner Solar System

1. To be precise, KBOs vary from the plane by up to 30 degrees.

2. A number of excellent references are available for those who want to know more, but I would like to recommend the Web site of NASA Jet Propulsion Laboratory (JPL) at Caltech (www.jpl.nasa.gov). JPL heads up NASA projects for the robotic exploration of space and maintains amazing, publicly available databases of missions and images. For an introduction to planetary data, see sse.jpl.nasa.gov/planets/ (JPL) and a similar site at Goddard Space Flight Center, nssdc.gsfc.nasa.gov/planetary/. Another excellent source of information is Thomas Watters, *Smithsonian Guide: Planets*

(New York: Macmillan, 1995). The book is now a bit dated, but it presents a great deal of useful information in a very clear and readable format.

3. Real greenhouses work exactly the same way. The green glass in the windows allows sunlight in but does not let heat back out again. This allows botanists to maintain tropical temperatures in nontropical regions of the world.

4. A number of excellent references have been written about Mars in the past decade. I cannot recommend a particular book or article, however, because of all the new and interesting information becoming available daily. I do suggest checking out the Web pages for JPL (www.jpl.nasa.gov) and Goddard Space Flight Center (www.gsfc.nasa.gov). A search of those sites for Mars will uncover the most recent information available.

5. Notable craters include Eagle Crater, Endurance Crater, and Victoria Crater, all located in Terra Meridiani around the intersection of the equator and the prime meridian.

6. See http://mars.jpl.nasa.gov.

7. Steroids are one of many PAHs used for signaling in terran organisms. PAHs have several attached rings of carbon. Their complex structure and unique decay patterns make them good markers for life—many are too complicated to arise without living organisms to process them. See Chapter 15 for more on steroids.

8. NASA's *Mars Climate Orbiter* and *Polar Lander* and JAXA's *Nozomi*.

9. Solar System nomenclature is confusing not only in English but in classical languages as well. Astronomers named the planets after Roman gods, but moon names come from a variety of sources. The Greek Deimos and Phobos were attendants of the Greek god Ares, a less popular version of the Roman war god Mars.

10. The Outer Solar System

1. Lovers is a euphemism for any number of conquests and affairs recorded in Greek and Roman mythology. Apparently, it was not too difficult to find sixty-three names. Notable exceptions are Amalthea and Helike, Jupiter's nurses when he was a baby, and Jocasta, the mother of the Greek hero Oedipus.

2. This very brief introduction can be supplemented by looking at information available on the Web, particularly at the *Galileo* mission site: www2.jpl.nasa.gov/galileo. I also recommend S. Vance, "Exploration and characterization of Europa," in L. J. Mix, J. C. Armstrong, A. M. Mandell, A. C. Mosier, J. Raymond, S. N. Raymond, F. J. Stewart, K. von Braun, and O. Zhaxybayeva, eds., "The Astrobiology

Primer: An Outline of General Knowledge—Version 1, 2006," *Astrobiology* 6(2006): 791–793.

3. Sulfur dioxide surrounds another jovian satellite, Io. The largest moons of Saturn and Neptune, Titan and Triton respectively, have thicker atmospheres composed of nitrogen (N_2) and methane (CH_4).

4. This distance is called the Roche limit.

5. In Greco-Roman mythology, the Titans preceded the Olympian gods as rulers. Many of the Titans were siblings or offspring of Saturn. The other moons have been named for Greek gods (Janus and Polydeuces) and nymphs (Pan, Calypso, and Daphnis). More recently named moons include Norse giants (Mundilfari, Narvi, Skathi, Suttungr, and Ymir) as well as figures from Gaulish (Albiorix, Erriapo, and Tarvos) and Inuit (Ijiraq, Kiviuq, Paaliaq, and Siarnaq) mythology.

6. Twenty-four moons are named for characters from Shakespeare. An additional three—Ariel, Belinda, and Umbriel—are characters from Alexander Pope's *Rape of the Lock.*

7. John Couch Adams (1819–1892, England) and Urbain Le Verrier (1811–1877, France).

8. Thalassa and Proteus were actually early names for sea gods, though Proteus eventually became incorporated into the court of Poseidon.

9. Kuiper belt objects (KBOs) are occasionally also called Edgeworth-Kuiper belt objects or trans-neptunian objects (TNOs).

10. These are *International Cometary Explorer* (renamed in 1991 from *International Sun-Earth Explorer* which launched 1986, NASA), *Vega 1* and *2* (1984, Soviet), *Sakigake* and *Suisei* (1985, JAXA), *Giotto* (1985, ESA), *Deep Space 1* (1998, NASA), *Stardust* (1999, NASA), and *Deep Impact* (2005, NASA). Fascinating discoveries have been made by visiting, shooting, landing on, and sampling comets. *Stardust* returned samples to Earth in January of 2006. Planned rendezvous include *Rosetta* (2004, ESA) and *Dawn* (2007, NASA).

11. Observations of Pluto have been made primarily by the Hubble Space Telescope and the Infrared Astronomical Satellite.

12. Charon was discovered in 1978 and named after the boatman in Greek myth who ferried the dead across the river Styx into the land of the dead.

13. Important papers include: M. E. Brown, M. A. van Dam, A. H. Bouchez, D. Le Mignant, R. D. Campbell, J. C. Y. Chin, A. Conrad, S. K. Hartman, E. M. Johansson, R. E. Lafon, D. L. Rabinowitz, P. J. Stomski, Jr., D. M. Summers, C. A. Trujillo, and P. L. Wizinowich, "Satellites of the largest Kuiper belt objects," *Astrophysical Journal* 639(2006): L43–L46 and M. E. Brown, C. A. Trujillo, and D. L. Rabinowitz,

"Discovery of a planet-sized object in the scattered Kuiper belt," *Astrophysical Journal* 635(2005): L97–L100. Further information can be found at Mike Brown's Web site: www.gps.caltech.edu/~mbrown/.

14. Quaoar is named for the Tongva creation god. The Tongva are a Native American people originating near Los Angeles. Orcus is another Roman name for the god of the underworld.

11. Extrasolar Planets

1. Planetary data was taken from the Extrasolar Planets Encyclopaedia Web site in August 2008. Check out exoplanet.eu for more, and more up-to-date information. A published catalog can be found at R. P. Butler, J. T. Wright, G. W. Marcy, D. A. Fischer, S. S. Vogt, C. G. Tinney, H. R. A. Jones, B. D. Carter, J. A. Johnson, C. McCarthy, and A. J. Penny, "Catalog of nearby exoplanets," *Astrophysical Journal* 646(2006): 505–522. It is available online at exoplanets.org.

2. D. W. Latham, R. P. Stefanik, T. Mazeh, M. Mayor, and G. Burki, "The unseen companion of HD114762—A probable brown dwarf," *Nature* 339(1989): 38–40.

3. M. Mayor and D. Queloz, "A Jupiter-mass companion to a solar-type star," *Nature* 378(1995): 355–359.

4. The observed mass is equal to the true mass multiplied by sin i, where i is the angle of inclination, the angle measured between the line of sight and the plane of the planet's orbit. Thus you will often see mass predictions listed as M sin i, the true mass times the correction for unknown inclination.

5. A. Wolszczan and D. A. Frail, "A planetary system around the millisecond pulsar PSR1257+12," *Nature* 355(1992): 145–147.

6. A. Hewish, S. J. Bell, J. D. Pilkington, P. F. Scott, and R. A. Collins, "Observation of a rapidly pulsating radio source," *Nature* 217(1968): 709–713.

7. S. E. Thorsett, Z. Arzoumanian, and J. H. Taylor, "PSR B1620-26—A binary radio pulsar with a planetary companion?" *Astrophysical Journal Letters* 412(1993): L33–L36.

8. R. Silvotti, S. Schuh, R. Janulis, J.-E. Solheim, S. Bernabei, R. Østensen, T. D. Oswalt, I. Bruni, R. Gualandi, A. Bonanno, G. Vauclair, M. Reed, C.-W. Chen, E. Leibowitz, M. Paparo, A. Baran, S. Charpinet, N. Dolez, S. Kawaler, D. Kurtz, P. Moskalik, R. Riddle, and S. Zola, "A giant planet orbiting the 'extreme horizontal branch' star V 391 Pegasi," *Nature* 449(2007): 189–191.

9. M. Konacki, G. Torres, D. D. Sasselov, and S. Jha, "A transiting extrasolar giant planet around the star OGLE-TR-10," *Astrophysical Journal* 624(2005): 372–377.

10. G. Chauvin, A. M. Lagrange, C. Dumas, B. Zuckerman, D. Mouillet, I. Song, J.-L. Beuzit, and P. Lowrance, "A giant planet candidate near a young brown dwarf," *Astronomy & Astrophysics* 425(2004): L29–L32.

11. B. W. Jones, P. N. Sleep, and D. R. Underwood, "Habitability of known exoplanetary systems based on measured stellar properties," *Astrophysical Journal* 649(2006): 1010–1019.

12. G. Schilling, "Habitable, but not much like home," *Science* 316(2007): 528.

12. Life and Time

1. Stephen Jay Gould, *Full House* (New York: Three Rivers Press, 1996), 176.

2. *Full House* explores this in much greater detail. Readers with a more fictional bent may be interested in reading Kurt Vonnegut's *Galapagos* (New York: Dial Press, 1999), in which he explores the possibility that intelligence may not be as important as we like to think.

3. The International Commission on Stratigraphy maintains a Web site with up-to-date geologic timescales. For the most recent estimates as well as a detailed diagram showing all of the eons, eras, and periods of Earth history, check it out at www.stratig-raphy.org.

4. ^{235}U, pronounced uranium-235, has 92 protons and 143 neutrons in its nucleus, totaling 235. ^{235}U is only one of several isotopes of uranium. All isotopes have the same number of protons, but each has a different number of neutrons. ^{238}U is the most abundant isotope (~99.3 percent), followed by ^{235}U (~0.7 percent) and ^{234}U (~0.0055 percent).

5. S. J. Mojzsis, G. Arrhenius, K. D. McKeegan, T. M. Harrison, A. P. Nutman, and C. R. L. Friend, "Evidence for life on Earth before 3,800 million years ago," *Nature* 384(1996): 55–59.

6. The Cambrian is the first period in the Phanerozoic Eon.

7. Technically, red algae fall into the group Rhodophyta. Plants, including green algae, are in the closely related group Viridiplantae.

8. The Cambrian explosion was a seemingly rapid appearance of most major groups of complex animals and a diversification of other organisms.

13. Making Cells from Scratch

1. Some physicists have postulated quantum strings to form a level of matter and force below the elementary particles. Despite beautiful mathematical properties, no direct evidence exists for such strings to date.

2. Kenneth Ford has written a wonderful introduction to quantum mechanics: *The Quantum World: Quantum Physics for Everyone* (Cambridge, MA: Harvard University Press, 2004).

3. A number of other composite particles have been discovered. They include lambdas, sigmas, omegas, pions, etas, and kaons. The specific properties of these particles are unimportant to the present discussion, but they demonstrate the hierarchical structure of matter at the most fundamental level.

4. We often hear about the twenty amino acids occurring in humans. In the laboratory, however, a larger number of amino acids can be used. Other amino acids can be found in other organisms and organic chemists have generated hundreds of types. A smaller number, perhaps thirty or forty, have proven useful.

5. For more details and the endosymbiotic theory, see the end of Chapter 16.

6. The Space Studies Board of the National Academy of Sciences released a report on this subject in 1999. It was entitled "Size Limits of Very Small Microorganisms; Proceedings of a Workshop" and can be found at www7.nationalacademies.org.

7. Archaea, which will be introduced in Chapter 17, also have only one prokaryotic cell and are also called prokaryotes.

14. Building Biospheres

1. Personally, I avoid rutabagas, but you get the idea.

2. Of course, even the exceptions have exceptions. A process called conjugation allows some bacteria to reshuffle their genes. For the most part, though, bacteria reproduce by dividing into two genetically identical daughter cells.

3. This is commonly known as the biological species concept and attributed to Ernst Mayr. It is currently the most popular definition among biologists. A fuller explanation of the subtlety of Mayr's position may be found in K. deQueiroz, "Ernst Mayr and the modern concept of species," *Proceedings of the National Academy of Sciences USA* 102(2005): 6600–6607.

4. In fact, even Darwin seems uncomfortable with the idea of species. To quote Elliott Sober: "Although Darwin called his most influential book *On the Origin of Species by Means of Natural Selection,* he often expressed doubts about the species concept. He says, 'No clear line of demarcation has as yet been drawn between species and subspecies' (p. 51). He then notes that he looks 'at the term species, as one arbitrarily given for the sake of convenience to a set of individuals closely resembling each other' (p. 52). Perhaps a less elegant but more apposite title for Darwin's book would have been *On the Unreality of Species as Shown by Natural Selection.*" Elliott Sober, *Philosophy of Biology* (Boulder, CO: Westview Press, 1993), p. 143.

5. C. M. Cavanaugh, S. L. Gardiner, M. L. Jones, H. W. Jannasch, and J. B. Waterbury, "Prokaryotic cells in the hydrothermal vent tube worm *Riftia pachyptila* Jones: Possible chemoautotrophic symbionts," *Science* 213(1981): 340–342.

15. Molecules

1. Right- and left-handed versions of the same molecule (called enantiomers) can be distinguished using a property called optical activity. Chemists begin with polarized light—light for which the electric field of all beams oscillates in a single plane. If you pass polarized light through a solution, perhaps a solution with a single enantiomer of glucose, the glucose will rotate the light one way or the other. If the light rotates to the right (clockwise), we call it dextrorotatory and assign the molecule a designation of "D." If the light rotates to the left (counter-clockwise), we call it levorotatory and assign the molecule an "L." The words come from the Latin for right *(dexter)* and left *(laevus)*. Unfortunately, no correlation exists between the handedness of a molecule (right or left, based on structure) and the optical rotation of that molecule (dextro- or levorotatory, based on rotation of polarized light). A right-handed sugar may or may not be dextrorotatory.

2. To be precise, starch involves $1,4$-α linkages and cellulose $1,4$-β linkages. The difference has to do with the chirality of the first carbon. Starch incorporates only α-glucose, whereas cellulose incorporates only β-glucose. Once the linkages form, the individual units remain stuck in one position or the other.

3. Lignin accounts for most of the remaining carbon in wood. Plants construct lignin by joining up units of chorismate, a ten-carbon molecule used in the construction of aromatic amino acids. Chorismate is a beautiful example of bottom-up construction. However the pathways evolved, chorismate has been appropriated by biochemical processes to form amino acids, essential vitamins, energy transfer molecules, and lignin.

4. The suborder Ruminantia includes cattle, deer, giraffes, goats, and a host of other grazing animals.

5. In other words, add $-NCHOCH_3$ to get N-acetyl-D-glucosamine.

6. Rare exceptions occur, including peptidoglycan and some natural antibiotics. They contain occasional D amino acids.

7. Some bacteria use the amino acid selenocysteine, which incorporates an atom of selenium. Methanogenic archaeans (see Chapter 16) utilize an amino acid called pyrrolysine, a highly modified lysine.

8. Technically, only cyanobacteria and chloroplasts (in plants) have chlorophyll. Other photosynthetic bacteria use a very similar molecule called bacteriochlorophyll.

9. The nucleosides are adenosine, cytidine, guanosine, thymadine, and uridine.

10. This applies for glycolysis and the tricarboxylic acid (TCA) cycle, the most efficient metabolic method known for converting glucose into energy. More information appears in the next chapter.

11. This applies for the Calvin-Benson cycle, which is used by plants.

12. Of course, other mechanisms exist for both processes, but these two are the most common.

13. J. D. Watson and F. H. C. Crick, "A structure for deoxyribose nucleic acid," *Nature* 171(1953): 737–738.

14. For comparison, consider the information stored on your home computer. Since there are four bases, each would take two bits to record. That makes four bases per byte. The hepatitis B virus would take 750 bytes, much smaller than the storage area for an icon. *Mycoplasma genitalium* would take up 150 kilobytes, the size of a text document. *Homo sapiens,* humans, would cover 750 megabytes, close to the amount of information stored on a modern compact disc. The *Amoeba genitalium* genome consumes a massive 167 gigabytes, or thirty-five DVDs.

15. The enteric bacterium *Escherichia coli* has fifty-two. The rat has eighty-two.

16. Triglycerides are also called triacylglycerols.

16. Metabolism

1. A quick and dirty introduction to metabolism in astrobiology can be found in L. J. Mix, G. J. Dick, and F. J. Stewart, "Redox Chemistry and Metabolic Diversity," in L. J. Mix, J. C. Armstrong, A. M. Mandell, A. C. Mosier, J. Raymond, S. N. Raymond, F. J. Stewart, K. von Braun, and O. Zhaxybayeva, eds., "The Astrobiology Primer: An Outline of General Knowledge—Version 1, 2006," *Astrobiology* 6(2006): 797–799. A more extensive description of these processes can be found in R. H. Garrett and C. M. Grisham, *Biochemistry* (Florence, KY: Brooks Cole, 2004).

2. Does this sound familiar? A more detailed explanation is given in Chapter 6.

3. The most common of these are H_2, H_2S, S^0, Fe(II), NH_4^+, and Mn(II).

4. Examples include CO_2, NO_3^-, and SO_4^{2-}.

5. More precisely, the incoming photon promotes the electron to a higher orbital, where it is more weakly attached to the molecule.

6. More efficient methods include the hydroxypropionate cycle, reverse tricarboxylic acid cycle, and the reverse acetyl-CoA cycle.

7. Of course, far more sunlight is available in the first few feet. The vast majority of photosynthetic organisms, both in number and mass, exist there.

8. Higher-level taxonomy (groups above the level of genus) can be confusing

within the bacteria, and confused. Microbiologists are still arguing about the best way to classify and what constitutes which level of organization. The Chloroflexi have also been called Chloroflexaceae or filamentous anoxygenic phototrophs (FAPs). The Chlorobi also go by Chlorobiaceae or green sulfur bacteria.

9. These go by Rhodospirillaceae and Chromatiaceae, despite the fact that the former appears to be a bad group (the members are not one another's closest relatives).

10. Free hydrogen in water usually comes in the ionic form (H^+), with no electrons to donate.

11. I think the following hypothesis is quite compelling. A sulfur lithotroph evolved into a sulfur phototroph and then into a sulfur-free phototroph. Current knowledge about phototrophy fits well with this idea, but until we know a great deal more, we can only speculate.

12. The TCA cycle is also known as the Krebs cycle or citric acid cycle.

13. FAD, like NAD+, resembles a standard dinucleotide. The related molecule flavin mononucleotide (FMN/FMNH) can also be used to store energy.

14. Actually, chloroplasts are a subset of a larger group called plastids. Most plastids of interest, however, are chloroplasts.

15. *Dehalococcoides ethenogenes* removes chlorine from tetrachloroethene. What makes this particularly interesting is that humans have only been producing tetrachloroethene for the last century, meaning that the relevant metabolic pathway must have arisen very recently.

16. *Rhodobacter capsulatus.*

17. The Tree of Life

1. *Summa Theologica,* book I, question 75.

2. Aristotle attributes the idea of evolution and change to Empedocles in *Physics* II.8. Plutarch attributes the idea of common ancestry between men and fish to Anaximander in *Symposiacs* VIII.8.

3. Blue-green algae are now called cyanobacteria.

4. Lynn Margulis and Karlene Schwartz wrote a wonderful book entitled *Five Kingdoms.* The book briefly introduces the history of taxonomy before launching into detailed descriptions of most major groups of organisms known today. I recommend it for anyone who wants a reference book that will help them understand biological diversity. Lynn Margulis and Karlene V. Schwartz, *Five Kingdoms: An Illustrated Guide to the Phyla of Life on Earth* (New York: W. H. Freeman, 1998).

5. Shared derived characters are formally called synapomorphies.

6. Some debate still exists over exactly how true this is and in what way it should be judged. As a rule, however, most biologists accept the premise in some form or another.

7. The phylogenies represented in Figures 17.2–17.4 show current estimates of diversity and relationships among groups based on SSU rRNA data. These trees should be considered back-of-the-envelope sketches, rather than definitive statements about the history of terrean life. Mitchell Sogin (eukaryotes) and Norman Pace (prokaryotes), both involved in astrobiology, have provided far more detailed studies.

8. The technical terms in genetics are transduction (transfer of DNA by virus) and gene transfer agents (GTAs, viruslike particles for spreading DNA).

9. *Leuconostoc oenos* plays a special role in wine production. Acetic acid bacteria cause wine to become vinegar when exposed to air. *Erwinia* and *Klebsiella* species appear to be important for coffee production as well as lactic- and acetic-acid bacteria. Similar bacteria contribute to the flavor of chocolate.

10. Offensive biological weapons research and development was banned by the Biological Weapons Convention of 1972. At present, 155 countries have agreed to the convention, including the United States, Russia, and China.

11. Just a warning: some geologists use the second spelling (Archaean) to refer to the eon. To my knowledge, no biologists use the first spelling (Archea) for the domain.

12. Technically, thermophiles prefer to grow between 45°C and 80°C; hyperthermophiles, "extreme heat lovers," grow above 80°C. Crenarchaeota tend to be hyperthermophiles.

13. I should mention that Bacteria and Eukarya both contain thermophiles and psychrophiles, though the examples from Archaea are the most extreme.

18. Exceptions

1. A more complete discussion of origin research can be found in F. J. Stewart, "Evolution of complexity," in L. J. Mix, J. C. Armstrong, A. M. Mandell, A. C. Mosier, J. Raymond, S. N. Raymond, F. J. Stewart, K. von Braun, and O. Zhaxybayeva, eds., "The Astrobiology Primer: An Outline of General Knowledge—Version 1, 2006," *Astrobiology* 6(2006): 768–771.

2. All plants, in fact, undergo alternating generations, wherein one generation is diploid and the next haploid. In this case, the term "gametes" tends to be limited to the one-celled stage in the life cycle just before meiosis. The whole haploid stage is

called the "gametophyte." Botanists have always considered the haploid generation of plants to be alive and organismal.

3. The words "human" and "individual" are problematic. Aristotle argued that life required a vegetative soul, whereas humanity required a rational soul. For him—and subsequently Augustine and Aquinas—life began at conception, human life at the quickening, some forty to eighty days later. Please don't read this as an argument for or against abortion. For the moment, it is simply a statement that almost no one thinks of eggs and sperm as alive, but perhaps we should reconsider what we mean by "alive."

4. The idea of a syndrome-based definition of life arose in a conversation I had with Carol Cleland (a philosopher at the University of Colorado) and Woody Sullivan (an astronomer at the University of Washington). Both are active members of the astrobiology community.

5. I suspect that life could be described as a stochastic process with a very large number of states.

19. Intelligence

1. One might take a look at Bruce Jakosky, *Science, Society, and the Search for Life in the Universe* (Tucson: University of Arizona Press, 2006). Jakosky is a prominent astrobiologist and physicist at the University of Colorado, Boulder. In his book he explores questions of how astrobiology interacts with culture.

2. *Compact Oxford English Dictionary of Current English* (New York: Oxford University Press, 2005).

3. This quote comes from Max Weber, "Science as a Vocation," in H. H. Gerth and C. Wright Mills, eds., *From Max Weber: Essays in Sociology* (New York: Oxford University Press, 1946), 129–156.

4. I put the cone names in quotation marks because the light at those wavelengths does not correspond exactly to those colors as we experience them.

5. S. I. Tomarev, P. Callaerts, L. Kos, R. Zinovieva, G. Halder, W. Gehring, and J. Piatigorsky, "Squid *Pax-6* and eye development," *Proceedings of the National Academy of Sciences of the USA* 94(1997): 2421–2426.

6. Photosynthesis began at least 3.2 billion years ago. Eyes developed within the last 500 million years.

7. For a recent article in this area, see J. M. Plotnik, F. B. M. de Waal, and D. Reiss, "Self-recognition in an Asian elephant," *Proceedings of the National Academy of Sciences USA* 103(2006): 17053–17057.

8. C. R. Currie, B. Wong, A. E. Stuart, T. R. Schultz, S. A. Rehner, U. G. Muel-ler, G.-H. Sung, J. W. Spatafora, and N. A. Straus, "Ancient tripartite coevolution in the attine ant-microbe symbiosis," *Science* 299(2003): 386–388.

20. The Story of Life

1. nai.arc.nasa.gov/roadmap/.

Acknowledgments

I would like to begin by thanking all of my colleagues in the international astrobiology and bioastronomy communities. They never fail to be informative, insightful, and entertaining. Special thanks to everyone who went out of their way to help me understand so very many things. I am indebted to Rosalind Grymes, David Morrison, Estelle Dodson, and the Astrobiology Primer Working Group. Without the primer, I would never have been able to write this book. Particular thanks go out to the people who helped me with the science in this book: Kasper von Braun (astronomy), John Armstrong (extrasolar planets), Shanti Rao (cosmology), and Jason Raymond (planetary science). Jim Kasting, Andy Knoll, Avi Mandell, and Tara Luke helped me find and use illustrations.

Daniel Conklin, Paul Stimers, and Amy Swanson read chapters and gave comments. Michael Fisher, Kate Brick, Dimitar Sasselov, David Morrison, and four other reviewers kindly provided advice for revision. Naomi Pierce and Michael Fisher were instrumental in seeing that the book was published. Finally, and ultimately, thank God.

Index

Neptune, 45, 149–150, 151; atmosphere, 113–114, 149, 293. *See also* Triton

neutron, 37, 106–107, 108, 170, 179–180, 306n6

neutron star, 108–109

niche, 167–169, 194

nitrogen: abundance/availability, 126–127, 151, 238; atmospheric gas, 55, 114, 127, 132, 146, 174, 316n3; building block for life, 76, 79–80, 208; chemistry, 81, 83, 207; isotope, 13. *See also* amine group

nucleic acid, 72–73, 85, 128, 185, 199, 211–220, 229, 233, 235, 266, 268; and taxonomy, 252, 254. *See also* DNA; plasmid; RNA; virus

nucleoid, 185

nucleolus, 217, 257

nucleotide, 55, 87–89, 199, 211–216, 218, 267

nucleus (atomic), 66, 83, 170, 180

nucleus (cellular), 184, 186, 217, 225, 251, 257

Oort cloud, 122, 152–153, 273

Opportunity, 133–136

organelle, 184, 186, 193–194, 219, 225, 242

organic chemistry, 27, 76

organism: definition, 188–190; size, 188

organotroph, 230, 239, 256

origin of life: conditions for, 120, 130; life from nonlife, 5, 59, 65, 187, 220, 268–271, 294; rooting the tree, 254, 262; timing on Earth, 173–174

osmotic potential, 85, 88–90, 233

oxidation, 85, 229

oxygen: abundance, 111, 114, 127–128, 144, 174–176, 181; biological production, 86–87, 194–195, 234–236; building block for life, 76, 78, 80, 199–201, 215, 224; chemistry, 53, 73–74, 211, 311nn8,9; electron dump, 230, 241; in electron transport chain, 239, 241; formation, 75, 102–105; heme, 81, 210, 211; as a poison, 175, 261; polarity in water molecule, 71. *See also* atmosphere; manganese wheel; oxygenic revolution; ozone layer

oxygenic revolution, 127–128, 171, 175–176, 234–236

ozone layer, 107–108, 109, 195, 196, 235

Pathfinder, 135

peptide bond, 207–208, 219

Phanerozoic eon, 170–171, 176–177, 318n6

phosphate group, 78, 80–81, 89–90, 212–215, 223–224, 233

phosphorus: abundance, 111; building block for life, 76, 78, 80; chemistry, 83. *See also* phosphate group

photon, 86, 97–98, 106, 109–110, 178; hitting Earth, 130, 166; in photosynthesis, 210, 226, 231–233; in vision, 276–277

photosynthesis, 62, 167, 174, 175–176, 226, 232–236, 240, 242, 276. *See also* Calvin-Benson cycle; chloroplast; phototroph

phototroph: chemistry, 81, 86–88, 90, 110, 231–237, 239; evolution, 175, 235–236, 238, 242; oxygenic, 175–176, 234–236; taxonomy, 251, 255–258, 261. *See also* chloroplast; oxygenic revolution; photosynthesis

phylogenetics, 189, 192, 243, 245, 251–253, 255, 258, 260

planet: definition, 94, 151; formation, 111–117; location for life, 92, 94, 117–121. *See also* giant planet; rocky planet

planetesimal, 115–116, 137, 150

plants: and gravity, 145; history, 176–177; metabolism, 86–87, 205, 210, 211, 220, 223, 234; reproduction, 192; soul, 246; taxonomy, 64, 185–186, 246, 250–255. *See also* Calvin-Benson cycle; endosymbiotic theory

plasmid, 182, 257–258, 266–267

plate tectonics, 36–37, 126–127, 173, 249

Pluto, 45, 150–152

polymer, 73, 181, 199

polypeptide, 182, 207–211, 219, 252–253, 268, 307n13. *See also* prion

pornography, definition of life, 25–26, 264

prion, 266

prokaryote, 185, 189; chromosomes, 217–218; metabolism, 220, 242–243; taxonomy, 250–251, 253–254, 257–262. *See also* Archaea; Bacteria

protein, 73, 181, 182–183, 199, 202, 206–211, 223, 225, 261, 266, 268; translation, 217–220

Proterozoic eon, 170–171, 175–176

protest, 218, 251, 254, 257, 259

proton, 37, 83, 88–90, 179–180, 306n6; in stellar fusion, 106–108. *See also* hydrogen, gradient; hydrogen, ion

psychrophile, 261
pulsar, 158–159, 286

radial velocity, 154–157
redox chemistry, 85–88, 89, 229, 231
redox potential, 231
red shift, 46, 156, 309n12
reduction, 85, 229
reductionism, 16–19, 33–35, 77, 196. *See also* emergence
replication, definition of life, 30–32
retrovirus, 220, 265
ribosome, 219–220, 252, 254
RNA, 181, 211–216, 219–221, 253; retrovirus, 264–265; small subunit, 219, 252–253, 262, 323n7. *See also* ribosome; RNA world
RNA world, 268, 307n13
rocky planet, 82, 92, 113–116, 119, 121, 124; extrasolar, 161–162, 164

Saturn, 141, 145–147, 148; atmosphere, 113–114, 145; moons, 75, 146–147
sentience, 276–280
SETI, 286–287
sex, 32, 176, 189–190, 191, 254–256, 270, 319n3
silicon: abundance, 114–115, 137, 143, 145, 146; fusion, 105–106; silicon-based life, 82–83
soul, 246–247, 278
species: definition of, 24, 65, 190–193, 250–251, 253; and evolution, 5, 37, 167–168, 176, 248–249
Spirit, 136
spontaneous generation, 5, 18, 186
starch, 73, 203–206
steroid, 170, 226, 315n7
subatomic particle, 17, 37, 179
sugar, 29, 182, 200–206, 212–213, 234; chiral, 55–56, 202, 269; energy storage, 175, 181. *See also* carbohydrate; glucose
sulfur: abundance, 126–127, 238; building block for life, 76, 78, 80; chemotrophy, 90–91, 168, 193, 199, 230, 238, 261, 322n11; formation, 75
supernova, 106–108, 158–159

symbiosis, 193–194, 205, 236. *See also* endosymbiotic theory
symmetry, 14, 39–51, 55

taxonomy, 191–193, 245–254, 259
terpene, 226–227
terrean, 3
thermophile, 261
Titan, 145–147, 153; atmosphere, 146–147, 153, 293, 316n3
transit photometry, 154, 159–160
tricarboxylic acid (TCA) cycle, 239–240, 242–243, 321n10
Triton, 149–150, 316n3
trophism, 230–232

uniformity, 39, 45–51
uranium: composition, 180; formation, 108; uranium-lead dating, 170–172
Uranus, 147–149; atmosphere, 113–114, 147–148, 293

Venus, 113, 124, 125–125, 127, 128; atmosphere, 125, 128–129
Viking, 135
virus, 187, 217, 219, 220, 258–259, 264–266, 271
visible light, 109–110, 130, 196
volatile, 114–115; getting volatiles to Earth, 118–119
Voyager, 142, 145, 148, 149, 273

water: alternative media for life, 73–75; availability, 115, 118–121, 124, 150, 152, 181, 293; chemistry, 27, 38, 55, 71–72, 86–87, 229, 308n8, 322n10; on Earth, 101, 121, 126–128, 173–174; electron source, 175, 235–236; on Europa, 142–144; in humans, 71, 80, 209; on Jupiter, 141; on Mars, 132–134; medium for life, 2, 71–76, 92, 168, 292–293; on Mercury, 124; on Titan, 146–147; tolerance for dehydration, 71; on Venus, 128. *See also* dehydration reaction; lipid
Woese, Carl, 252–253, 259